99-HHE-5UR

DEVELOPMENT OF MOLECULAR METHODS FOR DETECTION OF INFECTIOUS VIRUSES IN TREATED WASTEWATER

by:

Theresa L. Cromeans, Ph.D.
University of North Carolina, Chapel Hill, NC
and Division of Viral Hepatitis, CDC, Atlanta, GA

Mark D. Sobsey, Ph.D.
University of North Carolina, Chapel Hill, NC

2004

WERF

The Water Environment Research Foundation, a not-for-profit organization, funds and manages water quality research for its subscribers through a diverse public-private partnership between municipal utilities, corporations, academia, industry, and the federal government. WERF subscribers include municipal and regional water and wastewater utilities, industrial corporations, environmental engineering firms, and others that share a commitment to cost-effective water quality solutions. WERF is dedicated to advancing science and technology addressing water quality issues as they impact water resources, the atmosphere, the lands, and quality of life.

For more information, contact:
Water Environment Research Foundation
635 Slaters Lane, Suite 300
Alexandria, VA 22314-1177
Tel: (703) 684-2470
Fax: (703) 299-0742
www.werf.org
werf@werf.org

This report was co-published by the following organizations. For non-subscriber sales information, contact:

IWA Publishing
Alliance House, 12 Caxton Street
London SW1H 0QS, United Kingdom
Tel: +44 (0) 20 7654 5500
Fax: +44 (0) 20 7654 5555
www.iwapublishing.com
publications@iwap.co.uk

Library of Congress Catalog Card Number: 2004108727

IWAP ISBN: 1-84339-722-6

University of North Carolina and Division of Viral Hepatitis, CDC, Atlanta, GA

ACKNOWLEDGMENTS

We thank the project manager for communications and guidance and the project subcommittee members for insightful comments and scientifically excellent questions. We acknowledge and thank the Division of Viral Hepatitis at CDC for allowing much of the work to be completed on the premises. We thank Dr. Betty Robertson for encouragement, intellectual contributions, and support of work on the replication studies of HAV strains. We thank Dr. Ralph Baric for intellectual contributions to the conceptual design of the detection methods. We thank Angela S. Foust for execution of much of the experimental work.

We thank the Public Works Department, Water and Sewer Division of DeKalb County, Georgia and the Town of Morehead City Wastewater Treatment Plant for participation in the project. We thank Greg Lovelace for technical participation and for sample collection at Morehead City. The wastewater samples were critical to the final evaluation of the methods.

Report Preparation

Principal Investigator:
Theresa L. Cromeans, Ph.D.
University of North Carolina, Chapel Hill
and Division of Viral Hepatitis, CDC, Atlanta, GA

Mark D. Sobsey, Ph.D.
University of North Carolina, Chapel Hill

Project Team:
Angela S. Foust, M.A.
University of North Carolina and
Division of Viral Hepatitis, CDC, Atlanta, GA

Gregory Lovelace
University of North Carolina

Project Subcommittee

Mr. Akin Babatola, *ESD Laboratory*
Dr. Dean Cliver, *University of California Davis*
Dr. Roger Fujioka, *University of Hawaii*
Dr. Gary Sayler, *University of Tennessee*
Dr. Moyasar T. Yayha, *Environmental Microbiology Consulting Services*

Water Environment Research Foundation Staff

Director of Research: Daniel M. Woltering, Ph.D.
Project Manager: Lola Olabode, MPH

ABSTRACT AND BENEFITS

Abstract:

Enteroviruses, hepatitis A virus (HAV), and other enteric viruses can survive wastewater treatment processes, even after chlorination, and are found in the final effluents. These viruses can be detected by cell culture techniques with observations for cytopathic effect (CPE). Recently molecular detection of viral nucleic acids has been used. Most viruses found in wastewater are RNA viruses and reverse transcription-polymerase chain reaction (RT-PCR) is a rapid and sensitive method to detect these single-strand RNA enteric viruses. This methodology does not distinguish between infectious and non-infectious viruses. Viruses inactivated in the treatment process can be detected but do not pose a public health threat. Methods are needed to quickly distinguish the infectious viruses from inactivated viruses, both of which may be present in effluents. In this project we investigated the use of a method that combines cell culture and molecular detection. If a sample contains viruses that replicate in cell culture even without CPE, the proof of replication can be demonstrated by the detection of a replicative form (RF) in cell culture that is only present during replication of infectious RNA viruses. A negative sense strand of RNA is generated from the viral positive strand virus, and these two are found primarily bound in a replicative form. This RNA was specifically detected by RT-PCR, including the positive and negative strands. This assay was developed and explored for the detection of low levels of CVB3 and HAV laboratory strains first and then for low levels of wild-type enteroviruses isolated from sewage samples. As few as four infectious units of wild-type enteroviruses, contained in 2.0 ml of water concentrate prepared from 600 ml of treated wastewater effluent, could be detected within two days. Treated wastewater effluents were collected weekly over an 18-month period, and viruses were concentrated with a developed method. About 7% of the final effluent wastewater samples were found positive for infectious enterovirus with the developed RF method, although only 1% of the samples were positive by presence of cytopathic effect. This method avoids the use of two complete cycles of cell culture to detect infectious virus, while confirming infectivity with a molecular method.

Benefits:

- The developed method will save time and money in comparison with the current standard method for detection of infectious viruses in environmental samples including treated wastewater samples.
- This method will select only infectious viruses that have replicated in culture; no false positive selection of input virus can be detected in wastewater concentrates.
- The method provides the necessary level of detection of infectious viruses in treated wastewater.
- A cell line was selected that can replicate higher titers of hepatitis A virus from stools.
- Chlorine-inactivated viruses were not detected with this method although other molecular methods could detect these viruses.

Keywords: Enteroviruses, hepatitis A virus, wastewater effluent, virus concentration, disinfection, virus replication

TABLE OF CONTENTS

LIST OF TABLES

LIST OF FIGURES

LIST OF ACRONYMS

AG	analytical grade
AMV RT	Reverse Transcriptase from avian myeloblastosis virus
ATCC	American Type Culture Collection
BE	beef extract
BGM	Buffalo Green Monkey kidney (cell line)
BSA	bovine serum albumin
bp	base pair(s)
Caco-2	human colorectal adenocarcinoma (cell line)
CDC	Centers for Disease Control and Prevention
CI	confidence interval
CL	confidence limit
CPE	cytopathic effect
CVB3	coxsackie virus B3 strain
CVB4	coxsackie virus B4 strain
DNA	deoxyribonucleic acid
DNase	deoxyribonuclease
dNTP	deoxynucleotide
DPBS	Dulbecco's phosphate buffered saline
DPD	N,N-diethyl-p-phenylenediamine
EDTA	ethylenediamine tetraacetic acid
EMEM	essential modified Eagle's medium
FBS	fetal bovine serum
FRhK-4	fetal rhesus kidney (cell line)
HAV	hepatitis A virus
IC-RT-PCR	immunocapture reverse transcriptase polymerase chain reaction
IgG	immunoglobulin G
M	molar
MA-104	rhesus monkey kidney
MB	molecular biology
MEM	minimum essential medium
$MgCl_2$	magnesium chloride
MOI	multiplicity of infection, number of infectious viruses per cell
MW	molecular weight
nt	nucleotide
NTR	non-translated region
PBS	phosphate buffered saline
PC	phenol chloroform
PCR	polymerase chain reaction
PEG	polyethylene glycol
PFU	plaque forming unit(s)
PI	post inoculation (infection)
PLC-PRF5	human liver cell (hepatoma) line

pptn	precipitation
RF	replicative form
RI	replicative intermediate
RNA	ribonucleic acid
RNase	ribonuclease
RT-PCR	reverse transcription polymerase chain reaction
SDS	sodium dodecyl sulfate
Taq	*Thermus aquaticus*
Tris	tris (hydroxymethyl) aminomethane
UNC	University of North Carolina
U.S. EPA	United States Environmental Protection Agency
VP1	viral polypeptide 1
W/V	weight per volume

EXECUTIVE SUMMARY

ES.1 Background

Enteroviruses, hepatitis A virus (HAV), and other enteric viruses survive wastewater treatment processes, even after chlorination, and are found in final effluents. The modes of disposal of the effluent lead to the potential for exposure of human populations and risk of enteric viral infection and disease. This is most true if receiving waters or aquifers are used as drinking water sources. Application of treated wastewater to agricultural land or to golf courses is another source of exposure to susceptible populations. Infectious viruses are detected in wastewater and in receiving waters by cell culture techniques and by molecular detection of viral nucleic acids. This latter method may lead to detection of viruses that no longer represent a risk to human health. Therefore, cell culture or animal hosts are required to conclusively demonstrate virus infectivity. Cell culture with two passages is most commonly used, and this is an expensive lengthy process. Recently the detection of the viral positive strand genome in inoculated cell culture by reverse transcriptase-polymerase chain reaction (RT-PCR) has been used to detect infectivity of viruses from environmental sources. However, input inactivated virus if adsorbed could be detected with this method. An alternative to RT-PCR is the analysis of inoculated culture for the negative strand (complementary to the positive sense viral genome), which is produced only during viral replication. The negative and positive strands form a double-stranded structure, the replicative form (RF), present in cells infected with positive strand RNA viruses, which includes almost all viral pathogens found in wastewater.

ES.2 Project Objectives

- Develop a sensitive and specific assay for the detection of RF containing negative strands of viral RNA in inoculated cell cultures
- Evaluate sensitivity of the method for detection of very low numbers of infectious virus units in an inoculum containing laboratory strains of viruses
- Evaluate wild-type viruses for production of RF and molecular detection in selected cell lines, including stool-derived HAV and sewage isolated enteroviruses
- Evaluate kinetics of production of replicative forms in infected cells
- Collect and archive treated wastewater samples for detection of viruses when the method is developed—if necessary a method of concentration of water for archiving will be developed
- Evaluate the developed method for detection of virus after disinfection to demonstrate lack of detection of inactivated viruses with the method and therefore selective detection of infectious viruses
- Test all collected water concentrates to determine presence of infectious viruses with the newly developed method

ES.3 Research Approach

1. The laboratory strain of HAV was first evaluated for development of the method. With the addition of an adequate infectious dose to appropriate cell culture, the method for detection of the replicative forms of the virus in culture was developed. Although the viruses found in sewage may replicate more slowly, this model virus provided an excellent system to study the technical design of the research approach. The conditions and treatments necessary to produce a preparation of RF RNA from infected cells were defined with the use of the well-characterized system. Since the infected cell contains an excess of the positive strands once replication is in progress, these must be removed in order to detect only the negative strands that are in the double-stranded RNA with the positive strand forms. All of the enzyme treatments and RNA extraction methods were first defined with this system. The molecular method for detection of these forms was defined with the RT-PCR detection of both positive and negative strands in equal quantity in the RNA preparations from infected cells.

2. The method developed for HAV was applied and evaluated in CVB3 infected cells. This model enterovirus has been well characterized and represents a large proportion of viruses present in wastewater. The kinetics of appearance of the RF and stability over time were characterized in this system.

3. Available preparations containing never-cultured HAV were characterized for replication in selected cell lines. Kinetics of RF production with these viruses was studied to understand the potential for detection of wild-type HAV.

4. Viruses isolated from sewage and only passaged one time in cell culture were studied for evaluation of infectivity with the method and in the presence of wastewater concentrates. The time of appearance at low levels of infection, similar to that expected from wastewater concentrates, was evaluated after infection with the viruses. The stability of the RF as markers of infectious virus production was evaluated until the time point that cells detached from the flask.

5. An improved method for concentration of infectious viruses from treated wastewater effluent was developed. This method was used to concentrate and archive samples for evaluation of the presence of infectious viruses with the developed method for molecular detection of the RF RNA.

6. The archived samples of 68 different time points from two different types of treatment facilities were analyzed by inoculation of the standard cell cultures. After incubation for 2 weeks, the cells were harvested, and the RNA was extracted, prepared for RF detection, and analyzed by RT-PCR.

ES.4 Conclusions and Recommendations

The proposed molecular method for detection of infectious virus was developed and validated for evaluation in cells infected with a very high level of virus as well as with very low levels of virus. Very low levels of infectious virus would be expected in treated wastewater samples. The kinetics of RF production in cells infected with environmental virus isolates demonstrated that infectious virus could be detected within days of inoculation in the appropriate cell line. An incubation period of two weeks would detect all viruses detected by the standard method in four weeks. The low quantity of virus expected in treated wastewater can be detected. With the developed method, infectious virus and non-infectious virus can be distinguished. A chlorine-inactivated enterovirus and infectious enterovirus were easily and rapidly distinguished with the assay.

An improved method for concentration of treated wastewater was developed and used to prepare concentrates of treated wastewater effluent collected weekly during one year. Seven percent of samples tested with the newly developed molecular method contained infectious enteroviruses. Only one percent demonstrated infectivity by the standard method of analysis. Thus the newly developed molecular method provided increased sensitivity over the standard method. Any laboratory performing the standard method could do the first part of this assay, however the molecular analysis could not currently be performed in wastewater utilities. After performing the standard cell culture inoculations, samples could be processed at a central location.

One cell culture line was identified as most susceptible to wild-type hepatitis A virus (HAV) isolates. Isolation of infectious HAV from environmental samples has been difficult to impossible. A much higher quantity of inoculum of HAV was needed to produce an infection than was needed for enterovirus infection and detection with the RF detection method. The low level inoculum usually expected in environmental samples such as concentrated wastewater will contribute to the difficulty in isolation and detection of infectious HAV.

The molecular method developed in this project should be further evaluated with collections of samples known to contain a higher percentage of infectious enteroviruses. This molecular method for RF detection relies on replication in cell culture that may occur with or without apparent CPE. In this study only the BGM cell line was used for isolation and detection of enteroviruses. Not all enteroviruses replicate on BGM cells. Inoculation of other cell lines simultaneously with the test samples, followed by evaluation of all the inoculated cell lines for enterovirus RF production should provide a higher percentage of positive samples. This approach should involve no more labor or expense than the current standard method and could yield increased detection of infectious viruses in environmental samples.

This method cannot be used on-site in wastewater utilities at the present time. Collaboration with a public health or university laboratory, however, would make this feasible. The first part of the RF assay could be performed with the same equipment and facilities required for the standard method, and then the samples could be forwarded to a laboratory designed to complete the molecular analysis. Further development of the assay, and new kits that make these assays easier to perform, also could enhance the utility of the assay for advanced metropolitan wastewater or water treatment facilities.

CHAPTER 1.0

INTRODUCTION

Enteric viruses have been detected in environmental waters and potable water supplies (Gantzer et al., 1998; Green and Lewis, 1995; Divizia et al., 1998). Molecular detection methods have improved the sensitivity, specificity, and ease with which viruses are detected. However, methods such as polymerase chain reaction (PCR) are unable to distinguish infectious from non-infectious virus particles. Many enteric viruses that pose a human health risk through waterborne exposure contain single-stranded positive-sense RNA genomes such as enteroviruses, caliciviruses, astroviruses and hepatitis A and E viruses. The replication of these viruses at the cellular level has some common features. Through receptor binding to the cell surface, the virus is taken into the cell where the protein coat is removed (uncoating) and the input viral genomic RNA can be used to generate proteins necessary for replication. At the same time a negative strand is made complementary to the positive strand of the virus. From the negative strands, positive strands are generated, and these are subsequently packaged in the new viral capsids to form viral progeny. The double-stranded positive and negative strands are called the replicative form (RF). A simplified schematic of this process is shown in Figure 1-1.

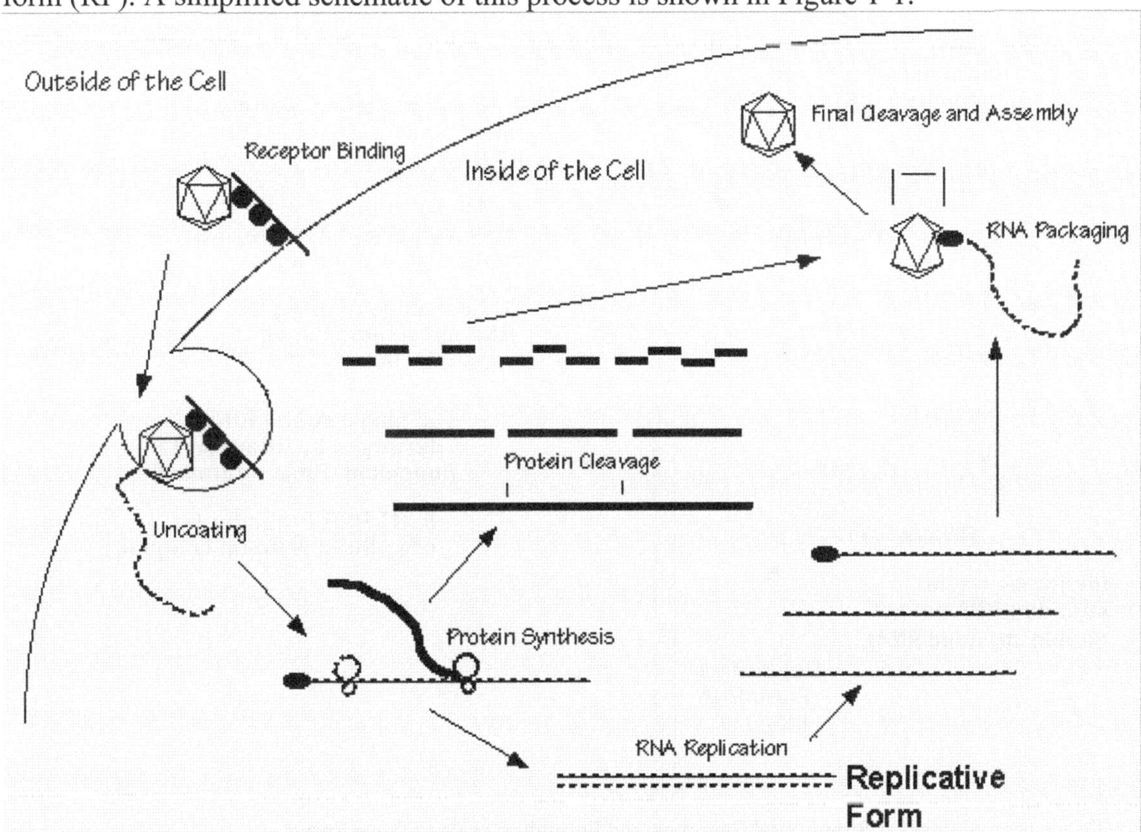

Figure 1-1. Basic Steps of Positive-sense, Single-strand RNA Viral Replication.

Replication of members of the Picornavirus family (including human enteroviruses, rhinoviruses, and HAV) has been investigated primarily through study of poliovirus replication in infected cell cultures. Poliovirus replication occurs on virus-induced membrane bound vesicles in the infected cell (for review see Johnson and Sarnow, 1995; Egger et al., 2002). All forms of replicating viral RNA and protein components of the viral capsids can be found on these membranes where viral replication occurs. The positive input strands of viral RNA generate proteins needed for formation of the negative strand. Positive strands made from the negative strands are used for either viral messenger RNA or new progeny virion RNA. As replication proceeds, the negative strands are in excess. The negative strands exist in infected cells almost exclusively as a partially single stranded, partially double stranded replicative intermediate (RI) RNA during production of positive strands or as the minor transient species of RF which is fully double stranded. Free negative strands have never been isolated. Treatment of RI RNA with specific RNase (RNase A and T1) degrades the single stranded, but not double stranded replicative form RNA (RF RNA) cores, representative of active transcriptional complexes in infected cells (Figure 1-2). Evidence of viral replication in cell culture, and therefore infectivity, is provided by detection of the negative strand specifically or by detection of these double-stranded forms containing one positive strand and one negative strand. Another important step in preparation of the RF for detection by RT-PCR, as described below, is DNase treatment. Any cellular DNA may yield non-specific priming during reverse transcription yielding a false positive result (Frech and Peterhans, 1994). To prepare the RF from infected cells, DNase treatment is followed by the specific RNase treatment.

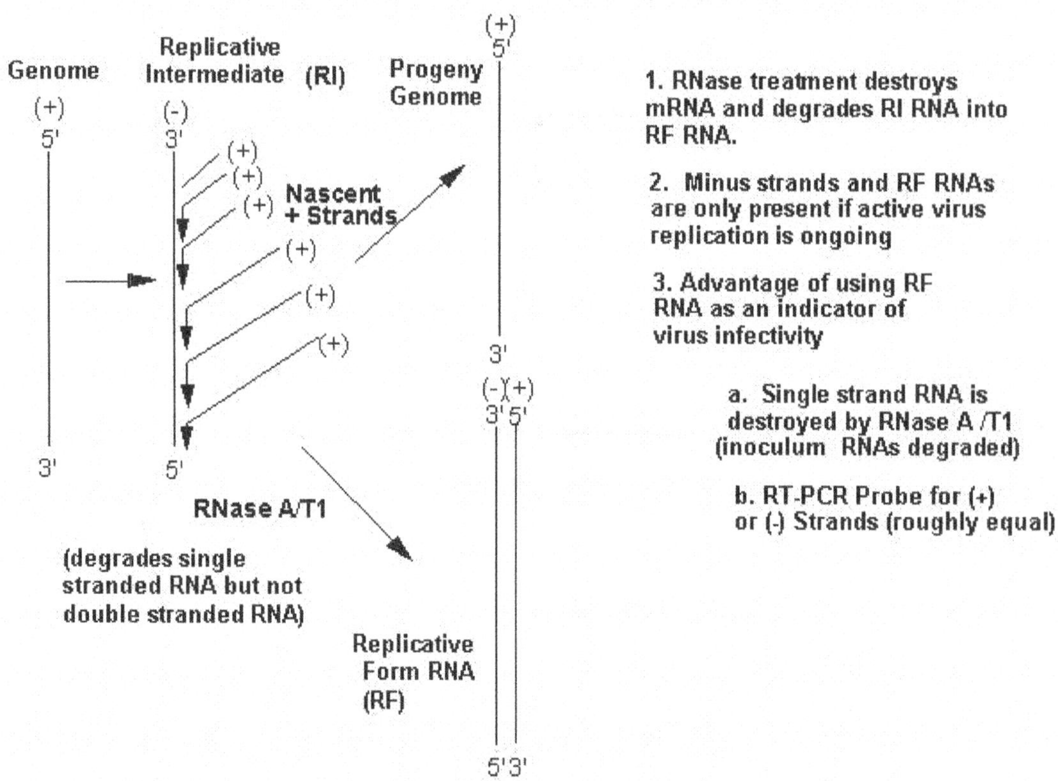

Figure 1-2. Replication of Positive Strand Viral Genomes and Formation of Replicative Forms.

Detection of the negative strand has been utilized for clinical assessment of replicating viruses in human tissues for hepatitis C and G viruses, coxsackie virus, and polioviruses (Andreoletti et al., 1997; Chaves et al., 1994; Craggs et al., 2001; Cunningham et al., 1990; Krawczynski et al., 1999; Lanford et al., 1994, 1995; Laskus et al., 1997a, 1997b, and 1998; Novak and Kirkegaard, 1991, Reetoo et al., 2000; Schmitz and Dotzauer, 1998). Detection of the positive genome strand in cell culture has been used to detect enteroviruses from environmental water samples (Chapron et al., 2000; Reynolds et al., 1996, 1998; Spinner and DiGiovanni, 2001). With this method, analysis of dilution series is required to confirm that detection by RT-PCR is not simply detection of the positive-strand genome of inoculated viruses. Any sporadic environmental sample concentrate containing a high level of viruses would be detected although the viruses could only be passively adsorbed to the cell culture without active replication. Detection of the replicative forms would confirm that the viruses are capable of replication and are infectious. Detection of the negative strand only could also confirm replication. However, there is a possibility of false positive results when relying only on the negative strand. Therefore, detection of the RF (positive and negative strands) after nuclease digestion of all single strands is a more definitive approach.

RT-PCR is a most sensitive method to detect RNA. Theoretically one nucleic acid copy can be detected with PCR, although this is not usually true for RNA detection. Reverse transcription of the RNA to make a cDNA, prior to the PCR, can be a limiting factor. RT-PCR also can be used to detect either of the strands (positive or negative) of RNA that make up the RF RNA in cells inoculated with infectious virus. The diagram in Figure 1-3 demonstrates an overview of the use of a unique enzyme that functions both as reverse transcriptase and as *Thermus aquaticus* (Taq) polymerase; thus one enzyme can complete the RT-PCR reaction in one tube. More commonly, two separate enzymes are used. With the primer P1 complementary to the positive strand, cDNA is generated in the presence of the necessary components at a relatively high temperature for increased specificity. After generation of the cDNA, the addition of the second primer P2 allows amplification of the newly made cDNA to a level that can be detected. In order to detect the negative strand, the P2 primer is added for the first reverse transcriptase step and then the P1 primer is added for amplification of cDNA. Separate reactions with a dilution series of a sample will yield a semi-quantitative analysis of the negative strand concentration and the positive strand concentration in the preparation. When RF preparations of infected cells are examined, equal quantities of the two strands are found. If only positive viral strands are present in the cell, they will be digested and not detected. In this project, we used this method to detect the RF in infected cells.

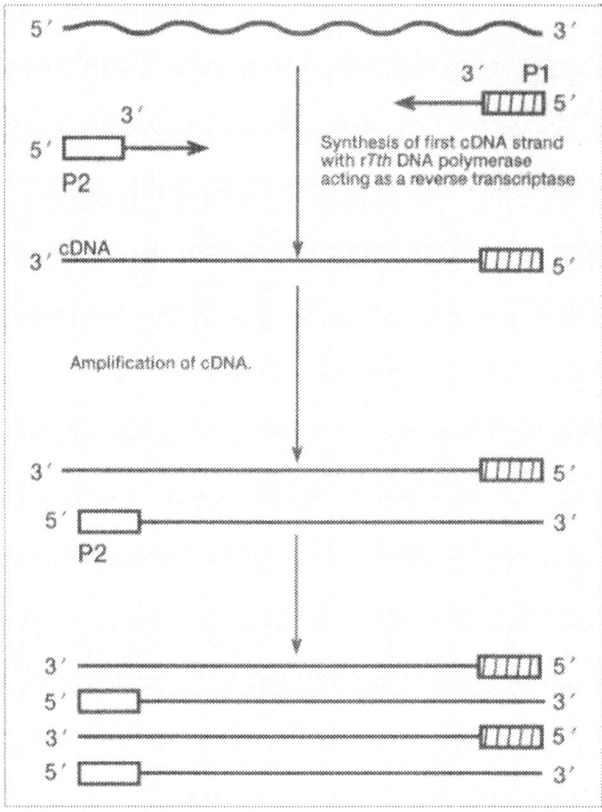

Figure 1-3. Schematic of RT-PCR with rTth Enzyme.

Knowledge of the kinetics of RF production in enterovirus-infected cells is limited primarily to knowledge of the process with infection at a very high multiplicity of infection (MOI) with poliovirus in HeLa cells and other supportive cells (Barton and Flanegan, 1997; Johnson and Sarnow, 1995). Little is known about the process in the infection of cells with a very low quantity of virus, typically found in most environmental samples, or with viruses other than poliovirus, which is currently unlikely to be found in the U.S. since oral polio vaccine use has stopped. This project explored the kinetics of production during infection with a model enterovirus, CVB3, and with enteroviruses immediately isolated from sewage. After establishing the method with high MOI, cells infected with a very low level of virus were evaluated for RF production over time.

Hepatitis A virus (HAV) is most stable in the environment and potentially more resistant to disinfection and environmental degradation than other enteric viruses (Cromeans et al., 2001). HAV is very slow to replicate in cell culture, does not cause cytopathic effect (CPE) and some isolates have been impossible to culture (Hollinger and Emerson, 2001). Evaluation of the potential infectivity of naturally occurring HAV in environmental samples has been almost impossible. Laboratory isolates causing CPE have been very useful in modeling experiments to evaluate environmental recovery methods and disinfection. More recently this CPE has been identified as apoptosis (Brack et al., 1998). A model cytopathic HAV rapidly replicating strain was used to establish the RF detection method for HAV infected cells. Several preparations of wild-type virus from stool samples were used to evaluate potential for replication and RF production in different cells lines. Since the wild-type viruses replicate without CPE, evaluation of RF production is a method to demonstrate replication. Detection of the negative strand/RF has been used to measure growth of HAV in murine cells (Takeda et al., 1996).

The developed method was evaluated for detection of replication of viruses potentially concentrated from secondary treated wastewater effluent from two treatment plants. In the process of evaluating coliphage as indicators, we found that the method we have used successfully previously for other food and water samples was not as successful as anticipated in recovery of viruses from the wastewater. We explored other similar methods and developed an improved method that was used to concentrate wastewater collected over the course of one year. All the archived sample concentrates were inoculated onto cell cultures, which were evaluated for the production of RF specific to any human enterovirus replication. The current standard method depends upon production of CPE for detection of infectious viruses. With the RF molecular detection method, infectious viruses that do not produce CPE may also be detected.

CHAPTER 2.0

MATERIALS AND METHODS

2.1 Wastewater Effluent Collection and Concentration

2.1.1 Wastewater Effluent Concentration for Coliphage Recovery Analysis and for Concentration of Archived Samples

Treated wastewater effluent samples from the wastewater treatment plant at Snapfinger, DeKalb County, GA were used for the coliphage recovery studies and for weekly sampling during the first three months of the study. Treatments at this plant include activated sludge treatment, denitrification, sand filtration, chlorination, and final dechlorination before release into the river. This is an advanced secondary wastewater treatment process including final chlorination and dechlorination before release into receiving water. Chlorine residuals before dechlorination of ranged from 0.42 to 0.74 on selected samples during the collection period. After dechlorination, chlorine residual was below the detectable level. The geometric mean of fecal coliforms in treated water was 400 per month during the time period samples were taken. Three one-liter samples were collected and processed for archives and an additional three liters for evaluation of coliphage recovery. The following parameters were evaluated in both preparations. After dissolving 0.3 M NaCl with stirring at room temperature in the samples, PEG was added and also dissolved before overnight stirring at 4°C. Polyethylene glycol (PEG) MW 8000 [molecular biology (MB) grade or analytical grade (AG), Sigma, St. Louis, MO] or MW 6000 (BDH, Poole, England) was added for final concentrations of 8, 10, or 12% (W/V). After overnight precipitation, solutions were aliquoted in 250 ml bottles, centrifuged at 10,000 x g for 30 minutes using a GSA rotor in a Sorvall RC-5B centrifuge, and supernatants were poured off. The bottles were rinsed with small volumes of Dulbecco's PBS (DPBS, Gibco, Grand Island, NY) to resuspend sedimented material on the sides and bottom of bottle (although no pellets were usually visible). Bovine serum albumin (BSA) at a final concentration of 0.1% was added to the concentrates. These concentrates and the supernatant were examined for somatic coliphages by standard double agar layer (DAL) plaque assay on *E. coli* CN13 host. Concentrates of 8–10 ml from the three liters for analysis of virus infectivity were archived at –20°C.

2.1.2 Sample Concentration for Analysis of Spiked HAV or CVB3 Recovery and Preparation of Archived Samples

Treated wastewater was collected from the Moorhead City, NC wastewater treatment plant starting in July 2001. The treatment process in this plant includes settling, trickling filter, and chlorination before release to provide a secondary treatment process including chlorination. Chlorination is regulated by an automatic flow meter and data is not kept on the amount, rather a control range of 0.2-2 mg/L is maintained in the final effluent. No historical data is available on the coliform range before chlorination. In addition to the routine collection of three liters of wastewater on a weekly basis, we collected additional liters of water for analysis of infectious virus recovery from wastewater spiked with laboratory strain HAV. Routine preparation of three liter samples was continued with the use of 8% PEG, until data were obtained demonstrating improved recovery of infectious HAV with the addition of 1.5% beef extract (BE), at which time the routine method was changed to include the addition of BE to the three liter samples as described below prior to addition of NaCl and PEG. This sample concentration method was used for one year with samples collected from Moorhead City, NC.

For experimental analysis of infectious HAV or CVB3 recovery, 500 ml samples of the wastewater were spiked with one ml containing 10^4, 10^5 or 10^6 PFU of virus. After mixing, 1.5% BE (Difco, Becton Dickinson, desiccated #211520) was added and dissolved prior to the addition of 0.3 M NaCl and the PEG of varying concentrations (8, 10, or 12% W/V). Experiments also were performed without the addition of 1.5% BE in which NaCl and PEG were added after spiking with virus. After overnight stirring at 4°C, samples were concentrated at 10,000 x g for 30 min. The concentrates were resuspended in 3.0 ml of Dulbecco's PBS and 0.1% BSA added. Concentrates were extracted once with chloroform. The interface was re-extracted with 2.0 ml of 3% BE and this aqueous phase was added to the first extract volume. Antibiotic/antimycotic (penicillin, streptomycin and amphoterycin B, Gibco, Grand Island, NY) was added to the final sample. This sample was assayed by plaque assay for recovery of HAV. A sample of the supernatant over the PEG pelleted material was also saved for plaque assay. Recovery was calculated based on final total PFU in the concentrate and the original total PFU spiked into the unconcentrated wastewater. Data were analyzed with the program GraphPad InStat version 3 (GraphPad Software, Inc.).

Archived samples for evaluation of infectious enteroviruses were prepared by different experimental methods through July 2000. Beginning in August 2000, all samples from Moorhead City were prepared with the selected method, which included addition of 1.5% BE prior to any other additions and use of 8% PEG (8000 MW, Sigma, St. Louis, MO, #2139, reagent grade) for precipitation as described for experimental samples. In the final analysis of the archived concentrates from both Snapfinger and Moorhead City, wastewater concentrates were thawed and extracted with an equal volume of chloroform. Sample and chloroform were shaken for 2.0 min and the aqueous phase was separated by centrifugation at 1200 x g for 20 min. Prior to inoculation of 2.0 ml samples on BGM cells, extracts in uncapped tubes were aerated for 1hr in a laminar flow hood to remove toxic effects of chloroform. BGM cells at passage level 128-135 were inoculated.

2.2 Virus Sources, Virus Propagation, Virus Preparations and Plaque Assays

2.2.1 Laboratory Propagated Viruses

HAV, HM-175 clone 24A, was propagated in FRhK-4 cells as previously described (Cromeans et al., 1989). A large stock of passage level 4 of this cloned stock has been prepared as previously described. Cells were harvested at one week post infection with 3 freeze-thaw cycles of cells and medium. This material was sonicated for 2 min in a water bath sonicator (Sonic dismembranator 60, Fisher Scientific) after which it was extracted with trichlorotrifluoroethane and the upper aqueous phase containing virus was frozen at –70°C. A titer of 4 x 10^7 PFU/ml was obtained. This titer is sufficient to infect a T-75 flask of FRhK-4 cells at a MOI of 4 PFU/cell, a high MOI yielding one-step growth curve conditions.

CVB3 (Nancy strain, obtained from ATCC) was passaged twice in BGM cells. The virus was harvested by freeze-thaw of cells three times and chloroform extraction. This stock had a titer of 2 x 10^7 PFU/ml in an assay of ten-fold dilutions on BGM cells, and 2.0 ml of stock was inoculated onto ~2 x 10^7 BGM cells in a T-75 flask. Experimental work with CVB3 was performed with BGM cells (CDC stock originally from Biowhittaker). All work with wild-type viruses (CVB4 and Echo 1) and analysis of field samples was performed with BGM cells (UNC, EPA source). Prior to evaluation of CVB3 in this cell line, a virus stock was prepared in BGM (EPA) cells.

Chlorine-disinfected CVB3 was prepared by standard methods. Sodium hypochlorite (6% household bleach) was used to make a 1 mg/L concentration of free chlorine and the concentration of free chlorine was measured by the standard DPD method. Virus stock was further purified and washed in 0.01 M phosphate buffer by centrifugal ultrafiltration (Centricon 100, Amicon Corp.) and dispersed by serial filtration through Tween 80-treated 0.2 µm and 0.08 µm porosity polycarbonate filters. Batch disinfection experiments used 1.7 x 10^3 PFU of CVB3 with 1 mg/L of free chlorine and a contact time of 10 minutes. The disinfected stock had no infectivity titer remaining when 0.2 ml aliquots were assayed in duplicate, and the same end-point RT-PCR titer as the control was obtained. Flasks were inoculated with the PFU determined prior to disinfection of stock virus; the RT-PCR titer inoculated was the same as the RT-PCR titer of the infectious virus.

Plaque assays of HAV and CVB3 were performed by inoculation of monolayers of the appropriate cells in 60 mm dishes with 0.2–0.25 ml of dilutions of the virus. BGM or FRhK-4 cells were prepared 2–3 days prior to inoculation and were just confluent. After one hr of adsorption of CVB3 or 90 min of HAV adsorption, 5.0 ml of MEM + 2% FBS + 1% agarose was added. Inoculated monolayers were incubated at 37°C for 2–3 days for CVB3 assay and 7–9 days for HAV assay. At the end of incubation, agarose was removed, and the monolayer stained with PBS containing formaldehyde and crystal violet as previously described (Cromeans et al., 1987).

2.2.2 Wild-type Virus Sources, Preparations and Characterization

Three wild-type HAV preparations from stools have been used to evaluate replication potential in several cell lines. HLD-2 HAV was obtained from a marmoset eight days after inoculation with a human virus isolate. An 18% (W/V) stool suspension was prepared in PBS as previously described and has been stored at –70°C for more than 10 years (McCaustland et al., 1982). This preparation has been evaluated for primate infectivity in Tamarins by inoculation of 10-fold dilutions of the HLD-2 stool preparation. One animal infectious dose was 10^{-6} /ml (CDC, unpublished). In preparation for cell culture infection, the 18% preparation was extracted with an equal volume of chloroform. The Q#1 stool preparation was obtained from an individual 2 weeks after onset of hepatitis A symptoms. This individual was infected in a small outbreak associated with a private well (DeSerres et al., 1999). A 20% stool suspension was prepared in Dulbecco's PBS by vortexing with glass beads for two min followed by centrifugation at 1000 x *g* to obtain the supernatant virus preparation. The virus preparation was extracted with an equal volume of trichlorotriflouroethane followed by extraction with an equal volume of chloroform. Both HLD-2 and Q#1 are genotype 1A HAV, the most common type found in the U.S. and worldwide. A third strain was used which represents genotype III HAV. The GA 76 strain was obtained from a chimpanzee nine days after inoculation with a human isolate (Khanna et al., 1992). To determine RT-PCR units contained in a given volume of the virus preparations, ten-fold dilutions of the preparations were analyzed. RNA was extracted from dilutions of these preparations with the QIAamp Viral RNA extraction kit (Qiagen, Valencia, CA), and end point titers determined by RT-PCR for positive-strand viral RNA (described below). Titers of HLD-2 and Q#1 were obtained by RT-PCR followed by second round PCR, however the GA76 titer was determined only by first round; the primers used for second round were not specific for GA76.

Enteroviruses were isolated from raw sewage collected after grit removal at an Atlanta, GA wastewater treatment plant. Three liters of raw sewage were collected and a polyethylene glycol precipitation was performed with 0.3 M NaCl and 8% PEG (8000 MW) as described above. Pellets were resuspended in Dulbecco's PBS, and the concentrate was extracted with an equal volume of chloroform. This process yielded ~50 ml of concentrate for isolation of enteroviruses. Sixty mm plates of BGM cells (UNC/ EPA) were inoculated with 0.3 ml of sewage extract. After 90 min, plates were overlaid with 5 ml of EMEM containing 0.5% agarose. Initially 30 plates were inoculated without any additive, and two plaque isolates were obtained in four days. In a subsequent isolation on 55 plates with the addition of 5 μg/ml of trypsin to the inoculum, nine plaques were isolated at 3 to 5 days PI. Plaques (5-10 mm in diameter) were visualized macroscopically and confirmed microscopically by the appearance of CPE. All of the agarose over a plaque was removed with 2.0 ml pipette tips and put into 2.0 ml of DPBS. After sonication, the material was chloroform extracted, the interface re-extracted with DPBS, and the upper aqueous layers were combined for analysis. Inoculation at 1:100 dilution of this material into T-25 cm^2 flask yielded CPE complete in 4-6 days for all 11 isolates. Amplification of RNA extracted from all the original plaque isolates yielded positives with the panenterovirus primers described below.

Two of the isolates were analyzed in the CDC enterovirus laboratory of Dr. Steve Oberste and identified by sequence analysis. Echovirus 1 and CVB4 were identified based on nucleotide sequence of 321 nt in the VP1 region of the genome (Oberste et al., 1999a, 1999b, 2000). Replication kinetics of these two viruses were analyzed. Infectivity titers of the original plaque isolates were determined by standard plaque assay on BGM cells. Echovirus 1 had a titer of 3.8×10^6 PFU/ml and CVB4 had a titer of 2.4×10^5 PFU/ml. These stocks were used in all evaluations of replication kinetics of the wild-type viruses.

2.3 RNA Extraction and RT-PCR Methods

Extraction and preparation of RNA was performed with different methods. The choice of method depended upon the source of RNA (viral preparations or infected cells) and type of RNA needed for analysis. Positive-strand viral RNA was extracted differently than replicative form RNA (containing the negative strand complementary to the positive strand). RF RNA was extracted directly from the infected cells with appropriate buffers for cell lysis. The positive strand viral RNA was extracted from a stock virus preparation, from infected cells in an aqueous preparation, or from supernatant culture medium of infected cells. Figure 2-1 shows the general scheme for preparation of the different types of RNA and the related assays that were performed.

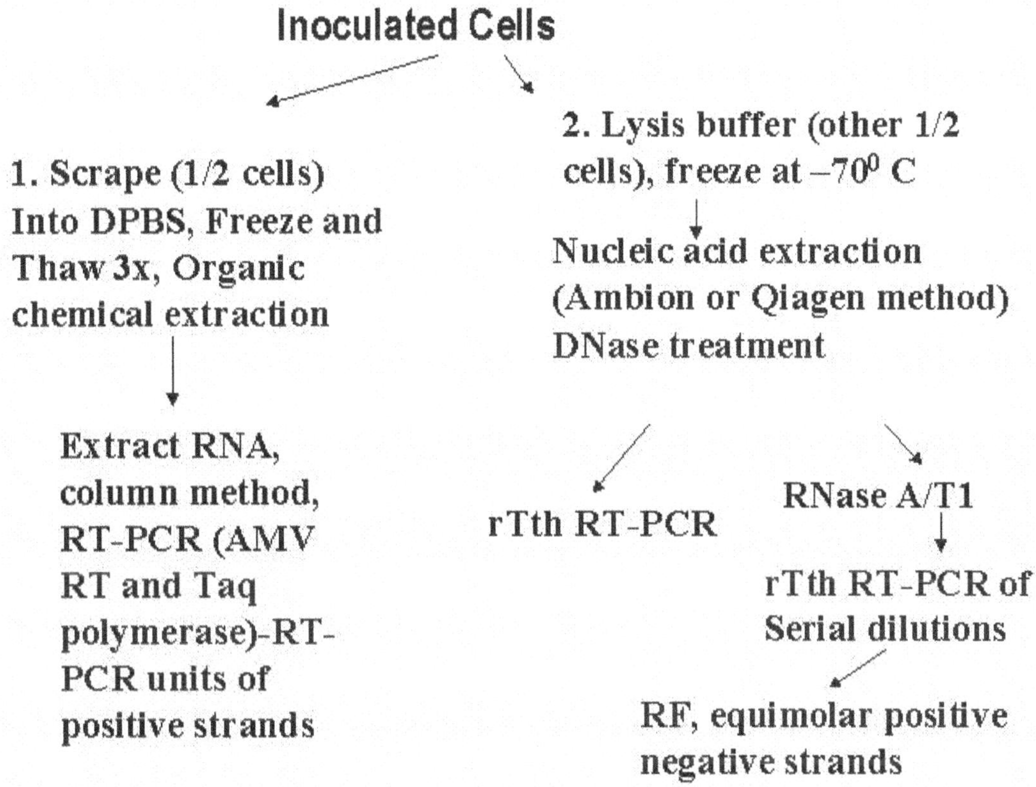

Figure 2-1. Preparation of RNA by Different Methods from Infected Cells.

2.3.1 Positive Strand Viral RNA Extraction and Detection

Viral RNA was extracted from dilutions of HAV or CVB3 stock virus with the QIAamp Viral RNA extraction kit (Qiagen, Valencia, CA), a method that combines the silica-gel-based membrane with microspin technology. The sample is lysed under denaturing conditions, adjusted to binding conditions and applied to spin columns, where RNA is bound and contaminants are washed away. RNA is eluted from the column with water. This extraction method was also used for detection of total RT-PCR viral units in infected cells harvested in an aqueous fraction. Infected cells were harvested in DPBS, frozen and thawed three times, and chloroform extracted. RNA was extracted from 10-fold dilutions of the cellular virus preparations with the QIAamp viral column method.

Oligonucleotide primers from the HAV genomic capsid region previously described for RT-PCR of HAV (Margolis and Nainan, 1990) were used. The 2167-2192 forward primer (2167-2192 relative to the genome sequence of HM-175 by Cohen and colleagues in 1987) and reverse primer (2390-2414 relative to the HM-175 sequence) yield an amplicon of 196 bp when visualized on 2% agarose gels with ethidium bromide staining. RT-PCR was performed similar to the previously described methods with some modifications (Cromeans et al., 1997). 25 µl of RNA was included in the RT reaction with AMV RT (Roche, Mannheim, Germany) for a final volume of 50 µl containing 5.0 mM $MgCl_2$, 200 ng of reverse primer, and 0.2 mM dNTP. After 60 min at 42°C, Taq DNA polymerase (Roche) and the forward primer were added to generate a final 100-µl reaction containing 2.5 mM $MgCl_2$ in a final volume of 100 µl. After 95°C for 2 min, 40 cycles of 95°C for 30 sec, 55°C for 1 min, and 72°C for 75 sec were performed followed by 7 min extension at 72°C. A second round of PCR was done with primers amplifying nucleotides 2195 to 2380 relative to the genome sequence of HM-175 of Cohen. The conditions of the second round PCR were the same as the first round, 30 cycles of the same cycling conditions were used.

The immunocapture RT-PCR method was performed similar to the published method (Cromeans et al., 1997) with some modifications. Microfuge tubes (0.65 ml) were coated overnight at room temperature with 0.65 µl of monoclonal antibody K2-4F2 (GL Limited, Melbourne, Australia) diluted in PBS. Experiments have been performed to determine that K2-4F2 is equally sensitive to the hyperimmune anti-HAV rabbit IgG used previously. Tubes were washed with PBS + 0.5% BSA (PBS/BSA) 2x before addition of virus dilutions in PBS, incubated 2 hr at 42°C, washed twice with PBS/BSA, and the RT buffer mix added in 43.5 µl. The tubes were heated at 95°C and frozen in a dry ice/methanol bath; after thawing, the RT mix of 6.5 µl was added and the 50 µl reaction incubated 1 hr at 42°C, then inactivated at 95°C, kept on ice for the addition of 50 µl of mix containing Taq polymerase, the forward primer and additional buffer were added and kept on ice until placed in the thermocycler heated to 60°C.

Enterovirus amplification of RNA was performed with primers previously described for the 5' NTR region of the genome, nucleotides 449-465 forward or 5' primer and 627-644 reverse or 3' primer (Schwab et al., 1995). All enterovirus experiments were performed with these primers. The same conditions described for the HAV RT-PCR were used. Thermocycling conditions were slightly different. After 95°C for 2 min, 40 cycles of 95°C for 45 sec, 55°C for 45 sec, and 70°C for 45 sec were performed, followed by 7 min extension at 70°C. Pan-enterovirus primers that identify all non-polio enteroviruses and used by CDC polio eradication initiative were also evaluated for wild-type virus and used for a second evaluation of wastewater concentrates that were positive for RF production with the UNC enterovirus primers. CDC primers generate an amplicon size of 114 base pairs with primers 5' – ACACGGACACCCAAAGTAGTCGGTTCC -3' and
5'- TCCGGCCCCTGAATGCGGCTAATCC – 3'.

2.3.2 Preparation and Detection of RF

Total RNA was isolated from infected cells and uninfected cells with Totally RNA (Ambion, Austin, TX) following instructions provided in the kit. Two and one-half ml of denaturation solution was added to ~ 1 x 10^7 cells at the indicated times PI. This solution was extracted with phenol:chloroform:isoamylalcohol, the aqueous extract was re-extracted with acid-phenol, and the final aqueous extract precipitated with isopropanol 2 hr or overnight at –20°C. This protocol was followed for all development and kinetic analysis experiments. Subsequently, the protocol was changed during analysis of wild-type enterovirus samples and for the field samples to use only an acid-phenol extraction followed by isopropanol precipitation.

The RNA pellet obtained by centrifugation at 10,000 x *g* was washed with 70% ethanol, and then DNase treated. RQ1 RNase-free DNase (Promega, Madison, WI) was used to degrade DNA that may remain in the RNA preparation. The DNase digestion was performed as described using the reaction buffer provided. Forty units of DNase was used for ~1 x 10^7 /cells or ~40 µg of RNA. After incubation at 37°C for 30 min, the RNA was treated with 20 mg/ml proteinase K and 0.5% SDS for 15 min at 37°C followed by re-extraction with phenol:chloroform:

isoamylalcohol, followed by isopropanol precipitation. An aliquot of this DNase-treated RNA sample was reserved for analysis, and the remaining sample was treated with RNase. Initial experiments evaluated different concentrations and types of RNase and all subsequent experiments included a 1:2000 dilution of RNase cocktail of A and T1 (Ambion, Austin, TX) containing RNase A at 500 U/ml and RNase T1 at 20,000 U/ml. RNase cocktail digestion was performed in a buffer containing 300 mM NaCl, 10 mM Tris pH 7.5, 5 mM EDTA. A buffer containing 700 mM NaCl, 10 mM Tris, 30 mM EDTA was used with RNase A (Sigma, St. Louis, MO) treatment. Reactions were incubated for 30 min at 30°C, treated with 20 mg/ml proteinase K and 0.5% SDS for 15 min at 37°C, and extracted with phenol:chloroform: isoamylalcohol, followed by isopropanol with 40 μg glycogen precipitation. Pellets containing the double-stranded RNA were resuspended in DNase- and RNase-free water or DNase- and RNase-free water + 0.1 mM EDTA. Preparations were either used immediately in RT-PCR reactions or frozen at –70°C for future use when 0.1 mM EDTA was included.

An alternative column method of extraction (RNase easy kit, Qiagen, Valencia, CA), was evaluated for isolation of RF fractions of infected cells. RNeasy commercial kit instructions were followed with the use of the QIAshredder spin column, which is purchased separately from the RNeasy column, and recommended for extraction of viral RNA from infected cells. Cells were harvested in lysis buffer provided with the kit and stored at –70°C until RNA extraction. After thawing, the lysate was loaded onto the QIAshredder spin column and the homogenized lysate of cells was collected. The QIAshredder was used with all trials of the RNeasy method. After addition of ethanol to the lysate obtained from the QIAshredder, the sample was applied to the RNeasy spin column for adsorption of RNA to the membrane. After washes with buffers provided, the RNA was eluted and ready for DNase and RNase treatment. A procedure for DNase treatment of RNA on the membrane using the RNase-free DNase Set Protocol from Qiagen was evaluated in conjunction with the column extraction method. DNase digestion is achieved while still bound to the silica-membrane with the special buffers provided for membrane digestion and has been optimized for use with the RNeasy extraction kit. The instructions with the DNase Protocol kit were followed. After elution of the DNase treated RNA from the columns, the RNase treatment was performed as described above. In other experiments RNA was eluted from the columns without DNase treatment on the column, and the standard DNase and RNase treatments described above were used.

The EZ rTth RNA PCR kit (Perkin Elmer, Branchburg, NJ) was used to evaluate all RNA extracted from infected cells. Briefly the components supplied including 5X buffer, rTth DNA polymerase and 25 mM $Mn(OAc)_2$ were used; dNTP (Roche, Indianapolis, IN) was used at a final concentration of 600 μM. Ten microliters of RNA were included in each reaction; controls included samples with no RNA. Control reactions with DNase- and RNase-treated RNA and no primers added were also performed. Primers were included at 200 ng/reaction. For positive strand reactions, the reverse primer was included in the RT reaction, and for RF detection, the forward sense primer was included in the RT reaction. Reverse transcription conditions were 60°C for 30 min, followed by 95°C for 2 min. PCR was performed with 40 cycles of: 94°C for 1 min, 60°C for 1 min, 72°C for 1 min, followed by 7 min extension at 72°C. Reaction products were examined on 2% agarose gels containing 5 μg/ml of ethidium bromide. Bands were visualized and photographed on a Stratagene Eagle Eye II system.

2.4 Experimental Infection of Cells with Virus Preparations or Wastewater Concentrates for Kinetic Studies and Detection of Replication

2.4.1 Infection of Cells with CVB3 and HAV HM-175

To analyze replication and RF production at different levels of infection or different MOI, cell monolayers were washed one time with DPBS and inoculated with the appropriate volume of inoculum containing the indicated PFU. T-25 cm^2 flasks were inoculated with 0.2 – 0.5 ml. Inoculated cells were incubated the indicated times at 37°C. At indicated harvest times, two types of cell harvests were performed. Cells were harvested by scraping into DPBS for analysis of infectious virus and of positive strand viral genome. This preparation was frozen and thawed three times, sonicated for one min, and chloroform extracted. The second harvest for RF was performed by lysis of the cells in the indicated buffers. In some experiments, separate flasks were harvested for each analysis. In other experiments, one flask was harvested for both analyses, by scraping one-half of the monolayer into DPBS, removing this sample, and removing the second half of the monolayer with cell lysis buffer for RNA extraction.

2.4.2 Infection of Cells with Wild-type HAV

Four different cell lines were evaluated for replication of wild-type HAV: FRhK-4 and MA-104 (Rhesus monkey kidney cell lines), Caco-2 (a human colorectal adenocarcinoma cell line), and PLC-PRF-5 (a human hepatoma cell line). Cells in T-25 cm^2 flasks containing ~2 x 10^6 cells were washed with DPBS and inoculated with 100 to 150 µl of the indicated stool preparations. After incubation at 37°C for 90 min, the inoculum was removed, and cells were washed with DPBS in initial inoculations. Inoculated cells were incubated at 37°C with EMEM + 2% FBS, and maintenance medium replaced weekly. At the indicated times PI, cells were prepared by two methods. After removal of the maintenance medium, one-half of the monolayer was scraped into 2.0 ml of DPBS. The other half of the monolayer was removed with lysis buffer for extraction of RNA and RF analysis. In some experiments all of the monolayer was removed into DPBS. The DPBS cell fraction was frozen and thawed three times, sonicated, and extracted with an equal volume of chloroform. This preparation was used to continue the passage of virus at the next level through the indicated number of passages. In those experiments where RF analysis was performed, the other half of the monolayer was processed as described for RNA extraction and preparation of RF.

In the viral adsorption studies performed in FRhK-4 cells, cells were harvested immediately after adsorption. Inoculum was removed, the monolayer washed with DPBS, and the cells harvested into DPBS by scraping the monolayer. Cells were prepared for viral genome RNA analysis as previously described. The cell preparation, the inoculum removed, and the wash were saved and processed by QIAamp viral method for measurement of viral positive-strand RNA. Percent adsorption was calculated from the PFU in the cell preparation/PFU inoculated times 100.

2.4.3 Analysis of Wastewater Concentrates for Infectious Virus

At 3-4 days after preparation, T-75 cm^2 flasks of BGM monolayers (UNC/EPA) passage levels 128-135 were inoculated with 2.0 ml of prepared wastewater concentrates (previously described). Medium was removed and the monolayer washed with 5.0 ml of DPBS, followed by addition of the wastewater concentrate. After one hr of incubation at 37°C, the inoculum was removed and EMEM 2% FBS was added to the monolayer. Control cells were mock inoculated with DPBS or MEM and maintained for each group of inoculated monolayers. Inoculated and control cells were observed daily during the 14 days of incubation. At 14 days after inoculation, monolayers were harvested in two fractions as previously described. One half of the monolayer was harvested in DPBS and frozen at –70°C for potential positive strand viral genome analysis, and the other half was removed with lysis buffer for preparation of RF RNA, as described previously.

CHAPTER 3.0

RESULTS AND DISCUSSION

3.1 Recovery of Naturally Occurring and Spiked Viruses by Wastewater Concentration Methods

3.1.1 Coliphage Recovery and Stability in Frozen Concentrates

Samples of chlorinated/dechlorinated wastewater effluent from Snapfinger Treatment Plant, DeKalb County, GA were concentrated with the 8% PEG method described. Concentrates were evaluated for recovery of somatic coliphages and compared to the somatic coliphage levels present in the initial effluent. Three liters of wastewater effluent contained 5-10 x 10^4 PFU somatic coliphages over five weekly collections, and recovery in PEG concentrates ranged from 2–21% (Table 3-1). This level of somatic coliphages indicated that a three-liter sample volume should be acceptable. The wastewater was evaluated for presence of residual chlorine, even though samples were taken after dechlorination. Residual chlorine could have inactivated viruses and yielded an inaccurate representation of viral infectivity after final treatment. No residual chlorine was found with the DPD colorimetric method, which could have explained the low level of coliphages.

Coliphage recovery and survival after freezing and thawing of concentrates was evaluated as an indicator of mammalian virus recovery after freezing and thawing of archived PEG concentrates. Frozen samples of wastewater effluent concentrates from the five weekly time collection points that yielded low recovery were analyzed for coliphage stability after freezing, storage at –20°C, and thawing. Results are shown in Table 3-1. The 95% confidence interval for each is shown in parentheses. Viruses were stable during freezing and thawing; only in one case was there a decline in number of coliphages after freezing (column 4).

Table 3-1. PFU/ml (95% CL*) Somatic Coliphages in Wastewater Effluent Concentrates.

Before freezing	4.5 (1.4-10)	5.4 (2.2-12)	8.1 (4.1-15)	14 (8-22)	1.8 (0.2-6.5)
After freezing	5.9 (2.9-10)	11 (7-17)	6.8 (3.8-11)	2.3 (0.7-5.2)	2.3 (0.7-5.2)

* 95% CL = lower and upper 95% confidence limit of reported value

The lower than expected recovery of coliphages could be due to the low initial concentration (about 3 PFU/ml based on average concentrations in unprocessed sewage effluent over a five week period), or the low recovery could be due to the exceptionally high quality of this treated effluent. In the previously published literature, reports of 90% recovery were obtained with high initial titers of 10^{11}–10^{13} PFU/ml in samples and only slightly lower recovery was reported with initially "low" titers of 10^5 PFU/ml (Yamamoto et al., 1970). These authors also observed a "plateau recovery of 50-70%" with the lower titers of phage.

In an effort to obtain better recoveries and have generally higher virus concentrations, wastewater samples were collected at a treatment plant in North Carolina previously documented to contain higher levels of coliphage in the effluent (Morehead City Municipal Wastewater Treatment Plant). A total six-liter volume of wastewater effluent was pooled and aliquoted in liter volumes. Processing of one set of samples was done at UNC and the other half shipped to Atlanta for processing in the CDC laboratory with the same methods. Coliphage analysis was performed at UNC on both sets of samples. Coliphage levels in the unprocessed samples were evaluated before and after shipment. Somatic coliphage concentration in this wastewater effluent was 9.6 x 10^4 PFU/ml before shipment and 5.7 x 10^4 PFU/ml after shipment. Table 3-2 shows the coliphage recovery efficiency in both laboratories with varying PEG concentration.

Table 3-2. Coliphage Recovery with Varying Levels of PEG with Intra-laboratory Evaluation.

PEG concentration	UNC Percent recovery	CDC Percent recovery
8%	25	24
10%	30	34
12%	32	38

The recovery is significantly higher in this secondary effluent from a facility using the trickling filter than it was from the wastewater effluents of an activated sludge plant. These results suggest that the higher initial coliphage concentration in a lower quality wastewater effluent sample yields higher virus recovery in the PEG concentrates obtained with reagent grade PEG and 0.3 M NaCl.

3.1.2 Recovery of HAV and CVB3 in Spiked Wastewater

Since the viruses to be analyzed are mammalian enteric viruses, we evaluated the recovery of HAV and CVB3 spiked into wastewater. Numerous experiments were performed in order to evaluate several potential parameters of the recovery method. A summary of all experiments examining infectious HM-175 HAV recovery from wastewater with varying conditions is shown in Table 3-3. Experiments were performed with different levels of HAV spiked in the wastewater; most experiments were performed with 10^5 to 10^6 PFU/ 500 ml. Determination of virus remaining in the supernatant was only made in the experiments with an inoculum of 10^5 or 10^6 PFU because with lower level spiking insufficient virus remains for direct measurement with the plaque assay.

Table 3-3. Spiked HM-175 HAV Recovery in Wastewater Concentrates and Supernatants.

Concentration Method	Mean Percent Recovery in Concentrate ± SE	Mean Percent Recovery in Supernatant ± SE
8% PEG	7 ± 1 (n=15)	34 ± 7 (n=11)
BE prior to 8% PEG	20 ± 4 (n=12)	20 ± 3 (n=10)
10% PEG	6 ± 2 (n=7)	21 ± 8 (n=5)
BE prior to 10% PEG	20 ± 3 (n=2)	11 ± 5 (n=2)

Data for recovery of infectious CVB3 are shown in Table 3-4. Because previous experimentation with HAV had suggested that virus remained in the supernatant after the first precipitation, the supernatant was re-precipitated with PEG under the same conditions as the first precipitation. Data in Table 3-4 suggest that a second precipitation is potentially less useful for virus recovery when it follows a first precipitation in which BE has been added to the sample.

Table 3-4. CVB3 Recovery from Spiked Wastewater.

Precipitation Method	Mean Percent Recovery After First Precipitation ± SE	Mean Percent Recovery After Second Precipitation ± SE
8% PEG	12 ± 1 (n=2)	12 ± 3 (n=2)
BE prior to 8% PEG	34 ± 12 (n=3)	10 ± 4 (n=3)

The parameters from the data presented in Tables 3-3 and 3-4 were analyzed by unpaired t tests with a two-tailed p value. Mean recovery with the parameters listed in column one and column two were compared (Table 3-5). Results in Table 3-5 indicate that precipitation with 8% PEG (either grade, MB or reagent) with BE added was the best condition for HAV and CVB3 recovery. Mean recoveries of HAV and CVB3 under the same conditions do not differ significantly. Recovery of HAV was not significantly better with 10% PEG when BE was added. The grade of PEG did not matter when BE was added to either. In summary, the data indicate that the use of 8% PEG reagent grade after the addition of BE was the best method for precipitation of the enteric viruses spiked into the treated wastewater effluent. Whether the viruses that may naturally occur are more likely to be recovered with the methods was not determined. The low levels expected to occur in chlorinated wastewater effluents make it difficult to evaluate within the scope of this project. Other investigators have found that extraction methods for sludge based on use of beef extract yielded higher recovery of viral titers by cell culture analysis and by nucleic acid detection (Monopoeho et al., 2001). In the study of Monopoeho and colleagues, naturally occurring viruses in sludge were quantified.

Table 3-5. Comparison of Mean Recoveries with Different Parameters.

Parameter 1	Parameter 2	*p* value (95% CI)
8% PEG, HAV	BE with 8% PEG, HAV	0.0048
10% PEG, HAV	BE with 10% PEG, HAV	0.0073
HAV, BE with 8%PEG	CVB3, BE with 8% PEG	0.1414
BE with 8% PEG, HAV	BE with 10% PEG, HAV	0.9413
MB grade 8% with BE, HAV	Reagent grade 8% with BE	0.3834
CVB3, one pptn. with 8% and BE	Two pptn. with 8% and BE	0.5271
CVB3, one pptn. with 8% PEG	Two pptn. with 8% PEG	0.0330

3.2 Development of RNA Treatment Methods for RF Preparation

3.2.1 RNA from HAV Infected Cells

HAV HM-175 (clone 24A) infected cells were examined for viral positive strand RNA and negative strand RNA after DNase treatment of a total RNA preparation from cells with Totally RNA (Ambion, Austin, TX). FRhK-4 cells infected at a MOI of 4 PFU/cell were harvested at two days PI. Dilutions of the RNA preparation were analyzed in an RT-PCR utilizing the rTth enzyme as described in Chapter 2.0. Results after DNase treatment are shown in Figure 3-1. RNA extracted with the QIAamp viral column method from HAV HM-175 virus stock preparations were also assayed (lanes 7,8 and 17,18). Amplicons were detected in 1000-fold higher dilution of RNA in the positive strand amplification than in the negative strand amplification. No amplicon products were found at the lowest dilutions (10^4 or 10^5) of RNA from uninfected cells. The stock viral RNA was only detected with the positive strand assay and not with the negative strand assay at the dilutions tested.

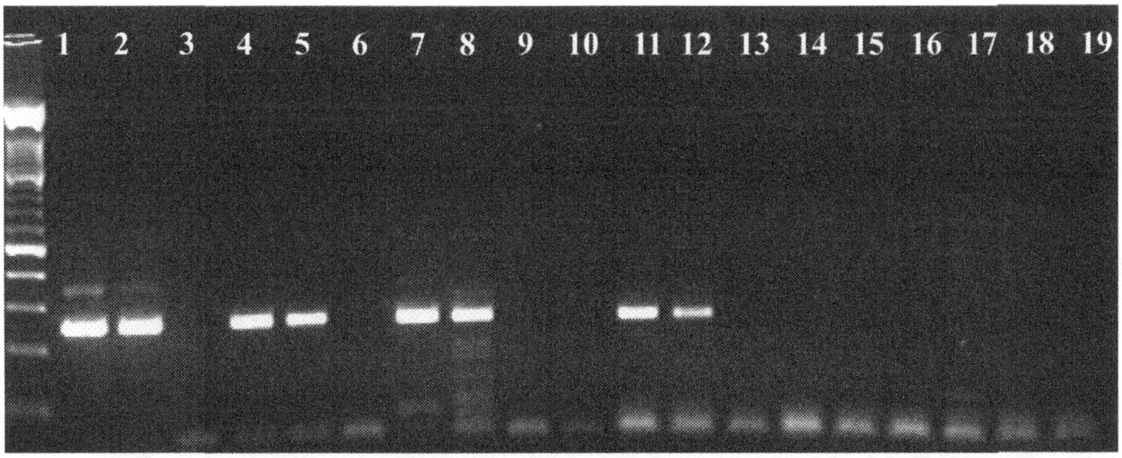

Figure 3-1. Detection of Positive and Negative Strands After DNase Treatment of Infected Cell RNA Extracts. The molecular weight marker in the left lane includes100 bp fragment markers and confirms the 247 bp HAV amplicon. Lanes 1-10 represent products generated in rTth RT-PCR with primers for positive strand RNA detection. Lanes 11-20 represent products generated with primers added for negative strand detection. Product in lanes 1-6 were generated from serial 10-fold dilutions (10^4 to 10^9) and in lanes 11-16 (10^4 to 10^9) of RNA extracted from infected cells; lanes 9, 10, and 19 were generated from uninfected cell RNA diluted to 10^4, 10^5, and 10^4, respectively. Lanes 7,8, and 17,18 amplicons were produced with RNA extracted from 10^4 (7, 17) and 10^5 (8, 18) dilutions of virus stock.

Digestion with RNase following DNase digestion was evaluated with RNase A/T1 at dilutions of 1:1000, 1:2000, and 1:4000 and with RNase A at 50 ng/ml. Results are shown in Figure 3-2.

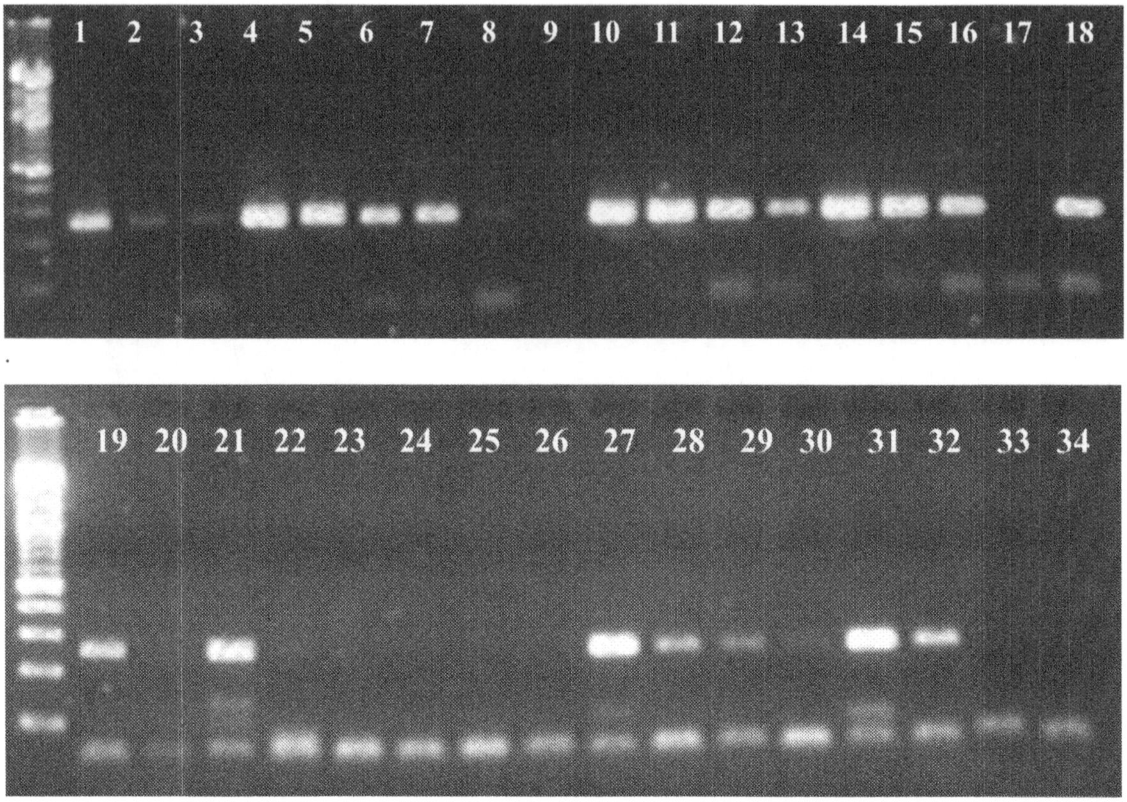

Figure 3-2. Comparison of Different RNase Concentrations for Detection of RF. In lanes 1-17 amplicons were generated with addition of primers for detection of positive strand RNA; in lanes 18-34 products were generated with primers added for detection of negative strand viral RNA. Amplicons in lanes 1-3 (RNA dilutions of $10^7 – 10^9$) and 18-20 (RNA dilutions of $10^4 – 10^6$) were generated with DNase only treated RNA from infected cells. Amplicons in lanes 4-6 (RNA dilutions of $10^3 – 10^5$) and 21-23 (RNA dilutions of $10^2 – 10^4$) were generated after treatment with 1:1000 RNase A/T1; lanes 7-9 (RNA dilutions of 10^2 - 10^4) and 24-26 (RNA dilutions of 10^2 - 10^4) after treatment with 1:2000 RNase A/T1; and lanes 10-13 ($10^2 – 10^5$) and 27-30 ($10^2 – 10^5$) after treatment with 1:4000 RNase A/T1. Amplicons in lanes 14-16 ($10^3 – 10^5$) and 31-33 ($10^2 – 10^4$) were generated from RNA treated with RNase A at 50ng/ul. Lane 34: control reaction with no RNA.

These results indicate that the 1:2000 dilution of RNase A/T1 is most effective in removal of single stands of positive strand RNA, including the viral RNA and the positive strands attached to the replicative form. RT-PCR analysis of the RNA digestion with 1:2000 dilution of RNase A/T1 was repeated to confirm this result. Another flask of cells infected at the same MOI was harvested at the same time, and RNA extracted and DNase treated. Results of a second trial are shown in Figure 3-3. The detection of positive strand RNA is reduced 100-1000 fold, and the detection of negative strands is also reduced by 10-fold after the RNase treatment. This result reflects the original excess of positive strands in comparison to the negative strands in infected cells, and also demonstrates the success of the RNase in reducing the positive-sense single strands. The 10-fold reduction of the negative strand detection may be variation in the assay or due to the reduction of non-specific amplification in the presence of the single-stranded RNA excess. Those strands remaining and detected in the assay are hybrids of negative sense and positive sense viral genomes, present as RF with the short negative positive strands removed, thus providing relatively equimolar amounts of each sense strand. The success of the rTth RT-PCR to distinguish the positive and negative strands is demonstrated by the comparison of lanes

1, 2 and 11, 12, where negative strands are not detected in the RNA from virus stock at these dilutions. Control reactions in which no primers were added yield no products at the lowest dilutions of RNA (results not shown).

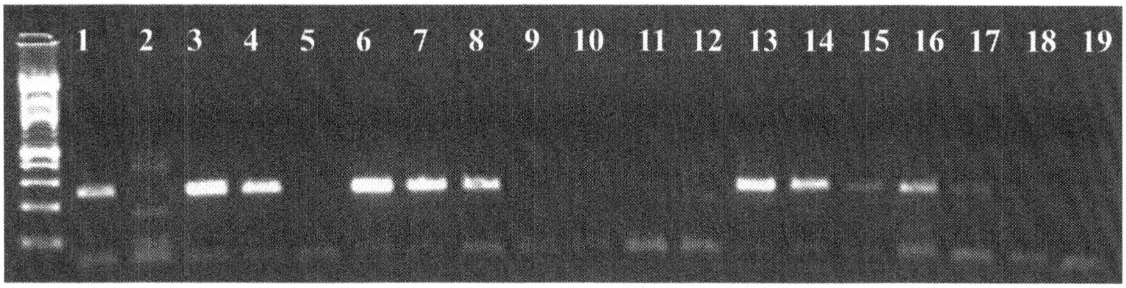

Figure 3-3. RNA titrations After DNase and Rnase Treatments. In lanes 1-10 amplicons were generated with primers for positive strand detection; in lanes 11-18 with primers for negative strand detection. Amplicons in lanes 1,2 and 11,12 were generated with RNA extracted from virus stock preparations (4 x 10^7 PFU/ml) by the column method, RNA dilutions of 10^6 and 10^7 were assayed; amplicons in lanes 3-5 (RNA dilutions 10^7 - 10^9) and 13-15 (RNA dilutions 10^4 –10^6) were generated from DNase-treated RNA. Amplicons in lanes 6-9 (RNA dilutions 10^1 - 10^4) and in lanes 16-18 (RNA dilutions 10^1 - 10^3) were generated from DNase and RNase A/T1 (1:2000 dilution).

The method was further evaluated with RNA preparations from FRhK-4 cells infected at lower MOI's with HM-175 HAV clone 24A. Cells were harvested for RNA preparation and evaluation at different time points. Results of these experiments are shown in Table 3-6.

Table 3-6. HAV RNA Detection in Cells infected with Laboratory Strain HAV (HM-175, clone 24A) at Varying MOI and Days PI—End-point Detection.

	DNase treated		DNase and RNase treated	
MOI (Days PI)	**Positive Strand**	**Negative Strand**	**Positive Strand**	**Negative Strand**
0.1 (2 da)	10^7	10^3	10^3	10^2
0.1 (4 da)	10^6	10^4	10^5	10^4
0.01 (5 da)	$\geq 10^8$	10^5	10^3	10^3
0.01 (7 da)	10^9	10^5	10^5	10^4

The data in Table 3-6 indicate that at lower MOI's and at different times after infection, the replicative forms can be identified after DNase and RNase treatments. Evaluation of RNA during the replication cycle is necessary with sampling at 8-12 hr intervals to determine the exact kinetics of appearance of the RF detectable in this assay. The replication kinetics of laboratory strain virus is of interest only to evaluate methodology and was not investigated in great detail. Kinetics of wild-type virus were investigated (see below). Other investigators have found that after appearance, the negative strands are present in the same quantity for weeks after infection with a slowly replicating laboratory strain of HAV (Schmitz and Dotzauer, 1998). Our data also indicate that the RF are stable over time or that any degredation of RF is followed by synthesis of RF such that a constant level can be maintained. Furthermore, the end-point analysis of RF (positive and negative strands after DNase and RNase treatment) suggests that there is an increase in quantity of RF between two and four days PI and five and seven days PI. At all these time points most cells remained attached although significant CPE was apparent.

3.2.2 Evaluation of RNA Preparation Methods for Detection of RF in CVB3 Infected BGM Cells and Determination of RF Kinetics

To evaluate the method developed for HAV infected cells for detection of the model enterovirus CVB3, BGM cells were infected with an MOI of 2 PFU/cell and harvested at 8 hr PI. Two flasks were infected, one flask was harvested for determination of virus titer and the other was harvested with lysis buffer for preparation of RNA with DNase and RNase treatment as previously described for HAV infected cells. The infectivity titer of the harvest from the duplicate flask of cells infected at high MOI was 2 x 10^7 PFU/ml. Results in Figure 3-4 demonstrate that after DNase and RNase treatment, the quantity of positive and negative strand RNA is equal, since a dilution of 10^3 was the end-point of amplicon detection in the assays for both positive and negative strands. At a relatively high MOI, the method successfully detects RF production in CVB3 infected BGM cells.

Result + + + + + + + + - + + + + + + + + -

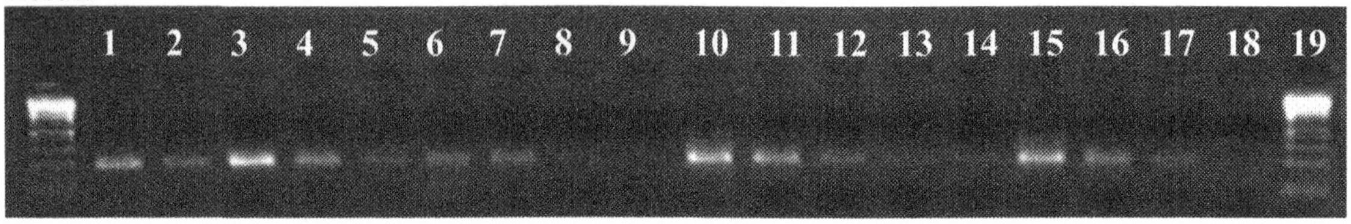

Figure 3-4. Detection of Replicative Forms in CVB3 Infected Cells with RT-PCR. Amplicons in lanes 1-9 were generated with primers added for positive strand RNA detection, and amplicons in lanes 10-18 were generated with primers added for negative stand RNA detection. In lanes 1-6 amplicons were produced from RNA dilutions of 10^3 to 10^7 of DNase treated RNA. In lanes 6-9, amplicons were produced from RNA dilutions of 10^1 - 10^4 of RNase A/T1 treated RNA. In lanes 10-14, amplicons were produced from RNA that had been DNase treated and diluted 10^1 - 10^5, and in lanes 15-18 amplicons were produced from RNA that had been RNase treated and tested at dilutions of 10^1 - 10^4. Interpretation of the signal on the gel is shown as result positive (+) or minus (-) across the top.

BGM cells infected at lower MOI (0.01-0.00001 PFU/cell) of CVB3 were analyzed for the presence of replicative forms at 24 hr PI. In a separate experiment, cells infected at 0.00001 MOI were analyzed at 48 hr PI. Results of these experiments are shown in Table 3-7. The end-point values were obtained by single tube assay of serial 10-fold dilutions. With decreasing MOI, decreasing quantities of positive strand RNA (measured after DNase treatment, second column in Table 3-7) were detected in infected cells. As expected, this result reflects a slower rate of viral replication with lower MOI. The quantity of positive strand RNA detectable after RNase treatment is three to four orders of magnitude lower than after only DNase treatment. After treatment with both enzymes, the quantities of RNA detectable by positive and replicative form/negative strand assays were approximately equal, indicating that only double stranded replicative forms were detected. Results in Table 3-7 demonstrate that an increased quantity of replicative forms (negative strands) was detected at 48 hr PI, as compared to 24 hr PI at the lowest MOI (equivalent to ~20 PFU/flask).

Table 3-7. Detection of Viral RNA in CVB3 Infected Cells.

	DNase treated		DNase and RNase treated	
MOI Hr PI	Positive strand	Negative Strand	Positive strand	Negative Strand
0.01 24 hr PI	10^8	10^4	10^5	10^4
0.0001 24 hr PI	10^7	10^3	10^3	10^3
0.00001 24 hr PI	10^6	10^3	$\geq 10^1$	10^2
0.00001 48 hr PI	10^7	10^4	10^4	10^4

A lower inoculum of virus would be expected from a wastewater concentrate sample, therefore a lower MOI, equivalent to 2 PFU/flask of cells was inoculated and examined for CVB3 replicative forms. RNA extracted from the infected cells was examined after both DNase and RNase treatments to detect replicative forms at 4, 8, 17, 24, 43, and 70 hr PI. No RF was detected until more than 40 hr PI and a similar quantity was found at 70 hr PI (data not shown).

To determine the time after infection of appearance of RF/negative strands, 4 PFU of CVB3 was inoculated per T-25 cm^2 flask of BGM cells. Two flasks were harvested at each time point; one to evaluate the yield of infectious virus and viral positive strand RNA, and the other to evaluate appearance of RF. Samples were taken at selected time points from 0 hr PI (end of adsorption period) until 44 hr PI. During the 44 hr incubation period, cells were observed for appearance of CPE. Before 28 hr PI most cells appeared normal. At 28 hr PI, about 5-10 % of the cells were rounded. At 32 hr PI, one area on the monolayer had 100% of cells rounded, which had the appearance of a large plaque on macroscopic observation of the flask. All rounded cells were attached. At 36 hr PI about 10-20% of the cells of the monolayer were rounded and the flask contained two of the larger foci or plaques with 50-100% of the cells rounded in the plaque area. At 44 hr PI 99% of the cells were rounded, although all cells remained attached.

At the indicated time PI, supernatant was removed from a flask, DPBS added to the cells and both fractions were frozen at –20°C. Both of these fractions were analyzed for infectious virus by plaque assay and the results are shown in Figure 3-5. The first cell-associated infectious CVB3 was detected at 8 hr PI, when no virus was detected in the supernatant fraction tested (0.5 ml of 5.0 ml total supernatant). The yield of infectious virus increased over time until 36 hr PI, when near maximum yield was obtained in the cellular and supernatant fractions. The largest increase in infectious virus production occurred between 8 and 20 hr PI.

Total RT-PCR units per flask were estimated from an end-point titration of the same preparation that was assayed in the plaque assay. Serial 10-fold dilutions were made of the chloroform-extracted cell-associated viral preparations, and RNA extracted from the dilutions with the QIAamp viral column method. The end-point titration, the volume assayed, and total volume of the flask preparation were used to calculate an estimated total RT-PCR units per flask. Results are shown in Figure 3-5. At the end of adsorption or 0 hr PI, 570 RT-PCR units were detected. At 0 time PI, the RT-PCR detectable viral units may represent input viral RNA contained in the inoculated virions or new viral (positive) sense RNA made soon after uncoating and initiation of viral RNA replication. Although some of the free viral RNA molecules may not be preserved in the freeze/thaw and chloroform extraction, a significant percentage may be present in the preparation. However, the variability of the plaque assay and the RT-PCR end-point detection may make it difficult to interpret the relationship at this particular time PI considering the very low levels present. Furthermore, the infection with 4 PFU per flask is only an estimate of the PFU contained in the dilution and based on the titer of the stock virus.

A large increase in cell-associated PFU and RT-PCR units occurred between 8 and 20 hr PI (Figure 3-5). At 20 hr PI, infectious virus is released from cells into the supernatant. The total PFU and total RT-PCR units are roughly equal from 28 hr PI to 44 hr PI. This approximate 1:1 ratio is different from the routine finding of 0.1- 0.01 PFU/RT-PCR units in the CVB3 stock virus preparations, which corresponds to a 10-100:1 particle to infectivity ratio. These ratios are similar to the expected ratio of non-infectious to infectious particles of 200:1 in picornavirus preparations (Racaniello, 2001). The RT-PCR is less efficient than a physical particle count.

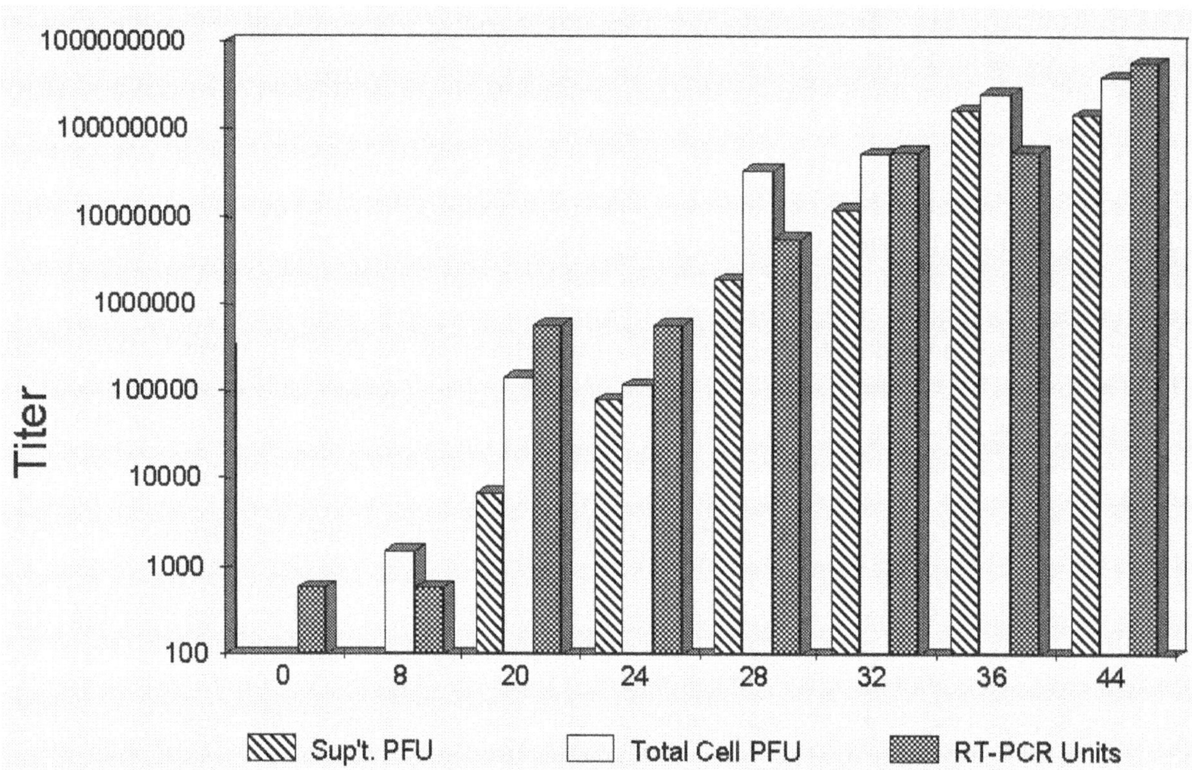

Figure 3-5. Comparison of Infectious CVB3 in an Extract of Infected BGM Cells and Supernatant with RT-PCR Units in the Extract of Infected Cells Over a Time Course of Hours After Infection at a Low MOI.

At the same times post infection, a second flask of cells was harvested with lysis buffer for preparation of RNA (Totally RNA, Ambion), nuclease treatments and analysis by RT-PCR with the rTth enzyme to detect RF/negative strands. Results of these analyses are shown in Table 3-8. In order to compare appearance of RF/negative strands with yield of infectious virus, the calculated PFU at each time point for the volume assayed in the RT-PCR is shown in the second column. Serial 10-fold dilutions of RNA preparations were assayed as previously described.

Positive and negative strand end-points after DNase treatment are shown in columns 3 and 4, and end-points after RNase treatment in columns 5 and 6. The RF/negative strands are detected after both DNase treatment followed by RNase treatment. These results are shown in the two columns on the right of Table 3-8. The end-point detection after assay for positive sense and negative sense are roughly equal, indicating that the double stranded replicative form is detected. As early as 8 hr PI, infectious virus, positive viral strands and RF can be detected. The RF end-points at this point, however, are not as strong based on examination of ethidium bromide-stained gels, since analysis of undiluted RNA is necessary and the specific band is weak with a slight haze of non-specific material stained by ethidium bromide. Definitive detection of RF was optimum at 20-24 hr PI in this experiment. It is clear that RF were detected before any significant CPE was detected; viral antigen would not be detectable with a simple enzyme immunoassay at this time point because such assays are several orders of magnitude less sensitive than either infectivity or RT-PCR assays. Also at the 24 hr time PI, RNase treated RNA was assayed with no primers in the RT reaction to check for background at dilutions of 10^1, 10^2 and 10^3, and no PCR products were obtained. In contrast, when specific primers were used in the RT reaction, specific product was obtained with a dilution as high as 10^4. This indicates that the product obtained is specific for RF detection.

Table 3-8. Comparison of CVB3 Infectious Units and Cellular CVB3 RNA, Positive Sense and Negative Sense, Over a Time Course After Infection with 4 PFU per T-25 cm² Flask of BGM Cells.

Hr PI	Calculated PFU[a] per volume assayed	DNase treated positive strand End-point	DNase treated negative strand End-point	RNase treated RF/positive strand End-point	RNase treated RF/negative strand End-point
0	<10	ND[b]	ND	ND	ND
8	7.5×10^1	10^1	ND	10^1	Undil
20	7.5×10^3	10^4	10^1	10^2	10^2
24	6.0×10^3	10^5	10^3	10^4	10^4
28	1.7×10^6	10^7	10^5	10^6	10^6
32	2.7×10^6	10^7	10^6	10^6	10^6
36	1.3×10^7	10^8	10^6	10^7	10^6
44	2.0×10^7	10^8	10^5	10^6	10^6

[a]The final volumes of each extract and the actual volume used in each type assay were used to calculate expected PFU. The total yield of RNA after DNase and RNase treatment was 200 µl from each flask and only 10 µl was assayed. To obtain calculated PFU, the total PFU obtained per flask was divided by 20.

[b]ND indicates not detected when 10 µl of the total 200 µl (from one T25 flask of cells) was evaluated in the rTth RT-PCR assay.

A relative comparison of RF kinetics and infectious units per volume (Figure 3-6) demonstrates that the RF detection is similar to detection of infectious virus. As early as 24 hr PI, even at the very low infectious input of 4 PFU, RF are detectable and are at peak production by 28 hr PI. RF can be detected until 44 hr PI when CPE is present in 99% of the cells. All of the data presented for detection of RF in CVB3 laboratory strain virus infected cells indicate that wild-type viruses that replicate in the BGM cells should be detectable with the method.

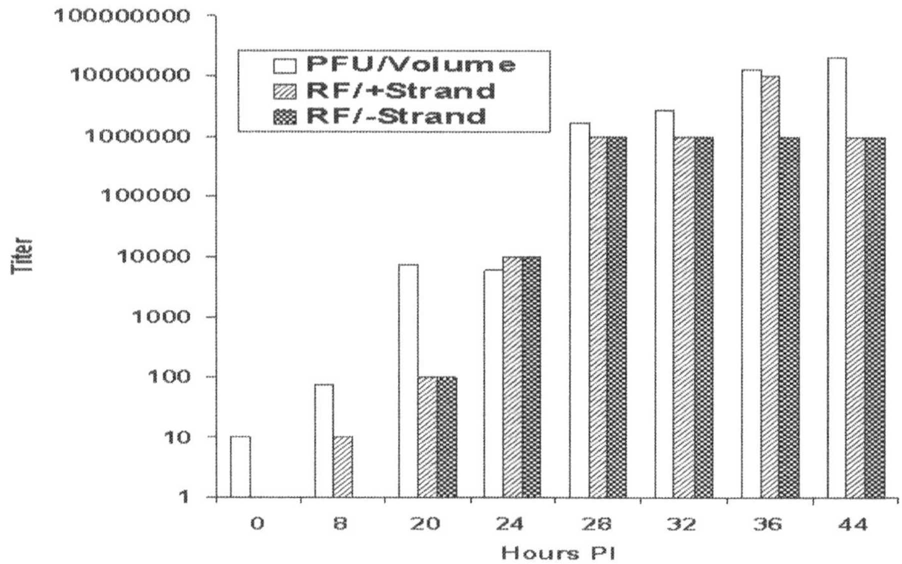

Figure 3-6. Kinetics of RF and Infectious CVB3.

There are no previous reports of enterovirus RF or negative strand production over time with a low MOI, such as the 4 PFU/flask or 0.000004 MOI. However, one study does examine poliovirus-infected cells at a MOI of 25 (Novak and Kirkegaard, 1991) at 2, 4, and 6 hr PI. The method we have used is similar to an RNase protection assay, and these investigators used a similar assay to detect negative strands with a radioactive probe, in contrast to the RT-PCR method that is more sensitive. They found an excess of 30-70:1 of positive strands in these cells with the highest excess at 4 hr PI. Early reports focused on the structural study of replicative intermediate and RF after extraction from poliovirus-infected cells at a MOI of 50 PFU/cell (Richards et al., 1984). Recent studies focus on the detailed molecular analysis of how the polar strands are generated and the location in the cell of the replication complex (Egger et al., 1996; Jurgens and Flanegan, 2003). These studies primarily use poliovirus with a high MOI and some are in cell-free *in vitro* systems (Barton and Flanegan, 1997; Barton et al., 1999; Teterina et al., 2001). Thus there are no data to compare with our kinetic results. It is not possible to know if RF declines in the initially infected cells and is then present in greater quantity in newly infected cells. However, the CVB3 kinetic study suggests that RF quantity is constant from 24 to 44 hr PI. Furthermore, the increase from 24 to 28 hr of RF is parallel with the increase in infectious virus. Both these facts suggest that RF (or the replicating intermediates from which they are derived) may remain in the infected cell rather than being degraded. The kinetic experiments demonstrate that the method is useful and that RF production is a reflection of production of infectious virus in cells infected at a low MOI.

All CVB3 experiments described thus far were performed with the phenol-chloroform method of RNA extraction from infected cells. Initial experiments with the RNeasy column method of RNA extract from high MOI HAV infected FRhK-4 cells suggested that this method yielded similar results to the phenol-chloroform extraction. Experiments were then performed to evaluate the column method for extraction of RNA and RF preparation from CVB3 infected BGM cells at a low level of infection. In this first comparison, the post column DNase treatment was used. Comparison of the two methods of RNA harvest from 8 to 36 hr PI is shown in Table 3-9. At 24 and 30 hr PI, the column method yielded 10-fold lower RF. This may or may not be significant given the variability of the end-point assay.

Table 3-9. Comparison of Methods for Detection of CVB3 RF in BGM Cells Infected at ~4 PFU/flask, End-point Detection of RF.

Hr PI	Phenol Chloroform method	Column method
8	10^1	10^1
20	10^4	10^{3-4}
24	10^5	10^4
30	10^5	10^4
36	10^6	10^6

A more detailed analysis was performed at the critical time points of 20 and 24 hr PI; each time point was evaluated with duplicate flasks to obtain more data for evaluation of the methods. In this experiment the DNase treatment was performed on the column. Results are shown in Table 3-10. When the four RF end-point values for each method are compared at the two different time periods, recovery is similar. However, somewhat higher end-points overall are obtained with the phenol-chloroform (P-C) method.

Table 3-10. Analysis of CVB3 RNA Extracted from Cells Infected with ~4 PFU/flask, Comparison of Phenol-chloroform (P-C) Extraction and Column Extraction with DNase Treatment on Column.

Hr PI and Method	DNase treated Positive strand	DNase treated Negative strand	RF/positive strand	RF/negative strand
20 P-C	10^6	10^5	10^4	10^5
20 P-C	10^6	10^5	10^5	10^5
20 Column	10^6	10^6	10^4	10^4
20 Column	10^7	10^5	10^4	10^4
24 P-C	10^7	10^6	10^5	10^6
24 P-C	10^7	10^6	10^5	10^6
24 Column	10^8	10^5	10^5	10^5
24 Column	10^8	10^6	10^4	10^4

Evaluation of chlorine-disinfected virus for production of RF in inoculated cells was performed. This is important, because many non-infectious viruses present in wastewater are disinfected, either by chlorine or UV disinfection. This analysis is important to prove the potential effectiveness of the method for discrimination of infectious and inactivated/non-infectious virus in treated wastewater effluent. The chlorine-disinfected inoculum did not produce infectious virus over the 35 hr incubation period, nor were RF produced (Table 3-11). In contrast, when the same quantity of untreated CVB3 was inoculated and evaluated by the same method at the time points, the expected virus genomes and RF were produced. The column method for preparation of RF RNA was also used in this experiment, and results are similar to that seen in Table 3-10. End-points at 20-24 hr PI are also slightly lower than seen in most experiments with the column method. The finding of 290 viral strand RT-PCR units in the cells inoculated with disinfected virus may be interpreted in several ways. First, virus could be adsorbed to the cells and remain detectable in the RT-PCR. The damaged virus may not be uncoated and the region of nucleic acid detected by the primers is still intact. Second, it is possible that there is a nonspecific product formed in the reaction of the total RNA from the cells with the primers in positive sense detection. This is likely based on our evaluation of uninfected cells. This fact also probably explains the detection of positive strands in the RF preparation but not negative strands. However, the value of 290 RT-PCR units is higher than usually observed in control cells. Both factors may contribute to the detection of positive strands in the total RNA preparation. Although RNA was amplified from cells inoculated with chlorinated virions, the results of this experiment show that RF detection is specific for infectious virus because no RF/negative strands were detected.

Table 3-11. Evaluation of Disinfected vs. Non-disinfected CVB3 for Detection of Total Virus RNA Production and RF in Cells Infected with 4PFU or Equivalent.

Time PI	Disinfected CVB3 inoculum-- RT-PCR units per flask	RF/positive strand end-point	RF/negative strand end-point	CVB3 untreated inoculum- RT-PCR units per flask	RF/positive strand end-point	RF/negative strand end-point
0 hr	2.9×10^2	10^1	Neg.	2.9×10^1	10^1	10^1
8 hr	NT	NT	NT	2.8×10^5	10^1	10^1
20 hr	2.9×10^2	10^1	Neg.	2.8×10^6	10^3	10^3
24 hr	2.9×10^2	10^1	Neg.	2.8×10^6	10^4	10^3
28 hr	NT	NT	NT	2.8×10^8	10^4	10^4
35 hr	2.9×10^2	Und.	Neg.	2.8×10^9	10^5	10^6

3.3 Replication of Wild-type HAV in Selected Cells Lines and Evaluation of RF Kinetics

Analysis of the potential for wild-type HAV replication was evaluated to predict the potential for replication of wild-type HAV in wastewater concentrates. Different cell lines have been evaluated with two different HAV strains in an attempt to determine the best conditions for detection of replication and production of RF/negative strands by wild-type HAV. A third strain of HAV was evaluated less extensively for replication in the cells lines.

To examine the potential for replication of HAV wild-type virus preparations, Q#1 and HLD- 2, FRhK-4 cells were infected with the prepared stool extracts. Monolayers of 2×10^6 cells were inoculated with the indicated RT-PCR units contained in 100 µl (Table 3-12). After adsorption for 90 min, the inoculum was removed, cells were washed with DPBS and Dulbecco's MEM + 2% FBS was added. Cells were refed at weekly intervals. Cells were harvested at four weeks PI by removing the supernatant and adding DPBS to cells. One half of the monolayer was scraped into the DPBS and removed. The other half of the monolayer was lysed with Ambion Totally RNA lysis buffer and stored at –70°C for analysis of RNA for RF. The cells scraped into DPBS were frozen and thawed three times and then extracted with trichlorotrifluoroethane to obtain a preparation for the next passage of virus and for titration by RT-PCR. Results of this analysis are shown in Table 3-12. After one month the quantity of virus in the cells is low. Virus detected in the supernatant would suggest some production and slow release since the cultures were refed weekly and this supernatant is only a one-week accumulation. However, it is possible that virus was adsorbed, not uncoated and slowly released from the cells.

Table 3-12. Analysis of Viral RNA One Month After Infection with Wild-type HAV.

HAV inoculum	Total input RT-PCR units	RT-PCR units total/cells	RT-PCR units total/supernatant
HLD-2	2.0×10^7	5.7×10^5	$\geq 7.0 \times 10^3$
Q#1 undiluted-A	4.3×10^4	5.7×10^2	5.0×10^1
Q#1 undiluted-B	4.3×10^4	5.7×10^2	5.0×10^1
Q#1 1:10	4.3×10^3	4.0×10^1	5.0×10^1

In order to evaluate the possibility that measurement of virus in the cells represents input virus adsorbed and/or uncoated after one month of incubation time, a virus adsorption study was done. Results are shown in Table 3-13. These results suggest that 1% or less of virus was adsorbed to the cells, 99-99.9% of virus was detected in the inoculum removed and/or in the cell wash following adsorption. The quantity of virus adsorbed to the cells is about equal to the quantity in the cells at one month after inoculation and culture, when potential range of detection of the assay is taken into account. The particle to infectivity ratio has been determined to be 1000-100:1 for many enteroviruses. Therefore when measured in RT-PCR units, only a fraction of the inoculated viruses would be adsorbed. Even when measured in cell culture infectivity, the particle to infectivity ratio for wild-type HAV has been described as 2.5×10^5 particles for one infectious unit in cell culture (Jansen et al., 1988). In contrast, cell-culture adapted virus stocks contained 58 particles for one infectious unit in this study, and 79:1 in a second study (Deng et al., 1994) in which a cytopathic strain of HM-175 18F was evaluated. When measured by plaque assay, 99.9% of a laboratory strain of HAV (Wheeler et al., 1986) was adsorbed. Measurement with genome detection instead of infectivity would yield different results. It would be of interest to perform this experiment. Thus, our adsorption results for wild-type HAV are not unexpected

Table 3-13. Adsorption of Wild-type HAV to FRhK-4 Cells Determined by Total RT-PCR Units per Fraction.

Wild-type HAV	Inoculum	Inoculum removed	Wash	Cells
HLD-2	2.0×10^7	1.4×10^5	$\geq 2.8 \times 10^7$	7.1×10^4
Q#1	4.3×10^4	2.1×10^4	7.1×10^3	7.1×10^2

The stability of one of the wild-type virus preparations at 37°C for one month was determined. The Q#1 stock undiluted and diluted 1:10 in DPBS was stored and end-point RT-PCR determined at the end of the month. The RT-PCR titer did not decrease (results not shown). This suggests that measurement of virus (positive strand RNA) in the cells without an increase in titer could be measurement of input virus, even after one month. Evaluation of HAV infectivity with only measurement of the positive viral strand could yield erroneous results, if virus is adsorbed but not replicating. A second passage of the virus is needed to confirm replication.

Initial analysis for RF in this first passage of wild-type virus was negative on first round RT-PCR. However, later analysis over the four-week incubation period and use of a secondary confirmation with nested PCR did reveal RF/negative strands in HLD-2 infected cells as described later (Table 3-19).

The virus preparations that were analyzed for viral RNA were also used to inoculate cells for the next virus passage. Results from serial passage of the HLD-2 and Q#1 are shown in Table 3-14.

Table 3-14. Analysis of Viral RNA at Passage Level One Through Three of Wild-type HAV.

HAV inoculum	Inoc. RT-PCR units for p.1	RT-PCR units total/cells p.1	RT-PCR units inoculated for p. 2	RT-PCR units Total /cells p.2	Inoc. for p.3	RT-PCR units Total/ cells p.3
HLD-2	2.0×10^7	5.7×10^5	flask A. 7×10^4	5.7×10^6	7×10^5	5.7×10^7
Q#1	4.3×10^4	5.7×10^2	flask A. 2×10^2	5.7×10^2	7×10^1	<10
			flask B. 2×10^2	5.7×10^2	7×10^1	7×10^1

Since only 1% or less of viral RT-PCR units was adsorbed in passage level one studies (Table 3-13), the increase in HAV RT-PCR units during the four weeks of each passage level could be even more than the apparent 80-fold increase, since the unadsorbed inoculum was removed. The adsorption kinetics of the second and third passage levels could be different from the adsorption of the first passage of virus. A higher percentage of viruses could be adsorbed than was adsorbed from the original stool suspension inoculum. At each passage level the HLD-2 titer increased 10-fold. Although the first passage and possibly second passage of Q#1 suggested potential replication, this was not sustained at passage level three. Whether this represents an initial replication that was not sustained cannot be determined from this data. Further work is needed to determine if this was an abortive replication that perhaps could be supported by altered culture methods.

During the second week of passage level two, morphological alterations were seen in FRhK-4 cells infected with HLD-2 and with Q#1. These changes were observed during passage level two and passage level three. Three independent observers verified the observation of an altered growth pattern and morphological changes. These changes did not develop into full cytopathic effect; there was no lysis of cells. Cells became rounded and granular in a distinct pattern within the monolayer. Interwoven chains of affected cells were seen throughout the monolayer. Discrete foci did not appear. Control cells never had this appearance, and control flasks were prepared in duplicate and observed over the four weeks of each passage. Overall, the percentage of cells affected roughly correlated with the increased quantity of HAV RNA in the preparation of HLD-2 infected cells. The Q#1-infected cells showed this effect dramatically, although low levels of viral RNA were found. At passage level two, 80% of the cells were affected. Analysis of the Q#1 preparations for other enteroviruses with panenterovirus primers by RT-PCR did not yield evidence of any enterovirus that could have caused the morphological changes. This effect was not sustained at passage level four. More investigation is needed to understand this phenomenon.

Evaluation of the replication potential of HLD-2 in different cell lines is presented in Table 3-15. Results of three separate trials in FRhK-4 cells are presented for comparison with initial trials in the other cell lines. Additional data on the HLD-2 replication in FRhK-4 cells are presented in another table below (Table 3-19). The total RT-PCR units recovered from a flask of infected Caco2 cells at passage level 1 was similar to that recovered from a flask of FRhK-4 cells infected at passage level 1 with HLD-2, which was the beginning of successful propagation of the virus. However, the second passage in Caco2 cells yielded a decreased titer. Passage level of the virus could influence replication potential; therefore passage was attempted in lower passage cells. The first passage in MA-104 cells was similar to that seen in FRhK-4 cells, however passage level two decreased significantly. The RT-PCR titer obtained from passage one of HLD-2 in PLC-PRF5 cells was 100-1000 fold lower than seen in the other type of cells, suggesting that replication may not have occurred. The titer after second passage in PLC-PRF5 cells decreased indicating lack of a sustained replication.

Table 3-15. Summary of Evaluation of Wild-type HAV, HLD-2, Viral RNA Production in Different Cell Lines.

Cell Line **Passage level**	**Virus passage number**[a]	**Total RT-PCR units per flask** [b]	**RF end-point**
FRhK-4, ~ 90	1A	5.7×10^{5}	10^{1-2}
85	1B	5.7×10^{4}	10^{1-2}
81	1C	2.8×10^{5}	10^{2}
Caco2, 34	1A	5.7×10^{5}	Negative
34	1B and C	5.7×10^{4}	NT[c]
24	2A, B and C	$<7 \times 10^{2}$	NT
43	2A, B and C	5.7×10	NT
PLC-PRF 5, 92	1	5.7×10^{2}	Negative
99	2	5.7×10^{1}	NT
MA-104, 155	1A	5.7×10^{5}	Negative
155	1B	5.7×10^{4}	Negative
171	2A	5.7×10^{2}	Negative
171	2B	1.4×10^{3}	Negative
82	2A	1.4×10^{3}	NT
82	2B	5.7×10^{2}	NT
182	3A and B	<57	NT

[a] A, B, and C represent different trials of the same experiment

[b] Calculated from end-point titration of a portion of the total volume of the flask of cells.

[c] Not tested

The Q#1 strain of HAV was also inoculated into the various cell lines. Results of these experiments are shown in Table 3-16. Passage level one and two of Q#1 in FRhK-4 cells suggested potential replication, however it was not maintained when passaged through level five. Whether this represented an abortive replication cannot be determined from our data. The infectivity of Q#1 is not known, whereas the infectivity of HLD-2 has been established in an animal model. The RT-PCR units per inoculum is at least 1000-fold lower than the HLD-2 inoculum. Two attempts to initiate an HLD-2 infection with a 1000-fold dilution of the stock virus were unsuccessful (data not shown). The RT-PCR titer of Q#1 from passage level one of Caco2 was 10-fold higher than was found in FRhK-4 cells, indicating potential replication. However, at passage two in Caco2, there was no increase in RT-PCR titer. Similar results were obtained with MA-104 cells. There was a decline in RT-PCR units obtained at the second passage level. Thus the Q#1 preparation was not successful in initiating infection in any cell lines that were examined under the conditions used. Investigators have reported culture of environmental HAV isolates with altered conditions of adsorption and culture.

Table 3-16. Evaluation of Wild-type HAV, Q#1, Viral RNA Production in Different Cell Lines.

Cell line Passage level	Virus passage[a] (Inoc. Dilution)	Total RT-PCR units per flask[b]	RF/negative strand assay
FRhK-4 ~90	1	2.8×10^2	NT[c]
~90	2	2.8×10^2	Neg
~90	3, 4 and 5	Neg	NT
Caco2 33	1A and B	2.8×10^3	Neg
37	2A and B	Neg	NT
PLC PRF5 96	1A and B	2.8×10^2	<20
96	1 C (1:3 inoc)	2.8×10^1	NT
96	2 A and B	Neg	NT
92	3 A	Neg	NT
MA-104 155	1	5.7×10^3	NT
171	2	5.7×10^1	NT
82	2	Neg	NT
182	3	Neg	NT

[a] A, B, and C represent different trials of the same experiment

[b] Calculated from end-point titration of a portion of the total volume of the flask of cells.

[c] Not tested

A third stock of wild type HAV (GA-76) was evaluated for replication in FRhK-4, PLC-PRF5, and Caco2. Results are shown in Table 3-17. The harvest at four weeks PI on first passage indicated potential replication in FRhK-4, however, the second passage declined in titer. The titer obtained at four weeks PI in Caco2 cells is below any titer seen with any successful passage level one of wild-type HAV HLD-2, and indicates lack of initiation of culture. PLC-PRF 5 cells appeared to have CPE and detached completely in one week; the cell lysate was assayed in RT-PCR and the flask contained 7 x 10^4 RT-PCR units. This could represent adsorbed inoculated virus. BGM cells (EPA origin) were also inoculated and cells deteriorated within one week, but recovered after refeeding. Since the same prepared inoculum of this virus did not cause any deleterious effects in FRhK-4 or Caco2, a common contaminant in the inoculum should not be the problem. No enteroviruses were detected with analysis of the stock and infected cells by RT-PCR with the panenterovirus primers; however, a primate virus not detected by these primers could be present and causing the apparent CPE in the cell lines. This HAV strain was not successfully cultured, although more attempts with the PLC-PRF 5 are needed when the interfering virus can be identified and removed from the inoculum. Further work is needed to determine if another virus is contained in the inoculum that may have caused CPE in two of the cell lines, and to determine if this virus isolate may replicate.

Table 3-17. Inoculation of Cell Culture with Genotype III Wild-type HAV.

Cells line/ Virus Passage		RT-PCR units/flask
FRhK-4	1	3 x 10^4
	2A	3 x 10^2
	2B	3 x 10^3
Caco2	1	<3 x 10^3
PLC PRF5	1 (1wk)	3 x 10^4

In summary, the only successful cell culture we have established of HLD-2 is in FRhK-4 cells. Although the RT-PCR titer of passage one of the inoculum can be detected in several cell lines, it does not represent establishment of replication. These titers may represent adsorbed and/or uncoated virus that remains stable during the four-week period or an abortive infection. We have demonstrated that virus incubated at 37°C for four weeks retains RT-PCR titer. Examination of cultures at weekly intervals of passage one could be useful in determining if RF are made and an abortive infection occurs. If these titers at passage one indicate that virus is adsorbed, then virus is potentially infectious in the susceptible host, such as humans.

The demonstration of successive passage of the HLD-2 in FRhK-4 cells was confirmed by the examination of the preparations by IC-RT-PCR in addition to the RT-PCR. The same preparations from infected cells that were used for the next passage inoculation were analyzed. In this method, virus is captured on antibody coated microfuge tubes and, after washes, RT-PCR is performed in the same tube. Only viruses with a capsid capable of immunological reaction should be captured and assayed with this method. Data in Table 3-18 demonstrate that IC-RT-PCR titers are 10-fold lower (passages one, two, and three) or equal to titer by RT-PCR (passage four and five). These results confirm that measurement of the increasing RT-PCR titers is a measure of viral production. The higher titer obtained by RT-PCR can be explained by some measurement of non-encapsidated viral RNA from the cells. Alternatively the IC-RT-PCR is not as sensitive, because not all viruses in solution may come in contact with the microfuge tube during the capture time. It is interesting that the two methods yield more similar titers at latter passages. Whether this is significant cannot be determined from this data.

Table 3-18. Comparison of RT-PCR and IC-RT-PCR Quantitation of Viral RNA in HLD-2 Iinfected Cells.

Passage level	RT-PCR yield	IC-RT-PCR yield
1	2.8×10^5	2×10^4
2A	2.8×10^6	2×10^5
3A	2.8×10^6	2×10^5
4A	2.8×10^6	2×10^6
5B	6.0×10^6	2×10^6

The complete analysis of passages of HLD-2 in FRhK-4 cells is shown in Table 3-19. Five successive passages of HLD-2 demonstrated the successful replication of virus. Detection of RF in early passage levels was difficult and not consistent with the first round of RT-PCR. Initially, at four weeks PI, the RF of early passage levels were not detected. RF production was seen definitively at passage level 5. Weekly analysis of the RF and total virus production was performed to determine kinetics and potential loss of RF during the four weeks of culture. To confirm the presence of RF, a second round of PCR was performed. Results in the two right-hand columns demonstrate that this second round of PCR analysis yielded confirmatory results for most end-points with some variation as expected from that determined by the first round of PCR. The highest RT-PCR titers were obtained at passage seven when the passage six virus was diluted 1:10 for inoculation. At week four of passage seven A, the titer is equal to that obtained with all well-established cell culture isolates of HAV, such as HM-175.

Table 3-19. Evaluation of Wild-type HLD-2 Replication for Total Genome RNA Production and RF Production at Weekly Intervals from Passage Level 1 to 7.

Passage[a] and week PI	Total Yield of RT-PCR units per flask	RF/positive strand evaluation End-point	RF/negative strand evaluation End-point	Nested of positive strand reaction	Nested of negative strand reaction
1.1 Wk4	5.7×10^5	10^2	Und	10^2	Und
1.2 Wk 1	5.7×10^5	Neg	Neg	10^1	10^1
Wk 2	5.7×10^5	NT	NT	10^1	10^1
Wk 3	5.7×10^5	10^2	10^2	Und	Neg.
Wk 4	5.7×10^4	10^2	10^1	10^2	10^1
1.3 Wk 1	2.8×10^5	NT	NT	10^1	10^2
Wk 2	1.4×10^4	NT	NT	Und.	Neg.
Wk 3	2.8×10^5	NT	NT	10^1	10^1
Wk 4	2.8×10^5	NT	NT	10^2	10^2
2A Wk 1	2.9×10^2	Neg	Neg	Neg	Und
Wk 2	2.9×10^3	10^1	Und	Neg	Und
Wk 3	1.4×10^4	10^1	10^1	10^1	10^1
Wk 4	2.9×10^4	Und	10^{-1}	Und	10^1
3A Wk 1	2.9×10^5	Neg	Neg	10^1	Und
Wk 2	2.9×10^5	10^1	10^2	10^1	10^2
Wk 3	1.4×10^6	10^2	10^2	10^2	10^2
Wk 4	2.9×10^6	10^3	10^2	10^3	10^2
4A Wk 1	2.9×10^5	Und.	10^{-1}	10^1	10^2
Wk 2	1.4×10^6	Und.	10^1	10^1	10^1
Wk 3	2.9×10^7	Und.	10^1	10^3	10^1
Wk 4	2.9×10^6	10^1	10^2	10^2	10^3
5A Wk 1	6.0×10^4	Und	Und	Und	Neg
Wk 2	6.0×10^5	10^3	10^{-4}	10^3	10^4
Wk 3	6.0×10^6	10^2	10^{-4}	10^2	10^4
Wk 4	6.0×10^7	10^2	10^{-3}	10^2	10^3
6A Wk 4	1.4×10^7	10^3	10^{-3}	10^3	10^3
6B Wk 4	2.9×10^6	10^3	10^{-2}	10^3	10^2
7A (1:10) 1	1.4×10^5	Neg	10^2	10^1	10^2
Wk 2	2.9×10^7	10^3	10^3	10^3	10^3
Wk 3	2.9×10^7	10^3	10^3	10^2	10^3
Wk 4	1.4×10^8	10^3	10^4	10^3	10^4
7B(1:10) 1	2.9×10^4	Neg	Neg	10^1	Und
Wk 2	1.4×10^5	10^2	10^2	10^2	10^2
Wk 3	2.9×10^5	10^2	10^2	10^2	$<10^2$
Wk 4	2.9×10^5	10^3	10^4	10^1	Neg

[a]Passage notation 1.1, 1.2 and 1.3 refers to separate attempts at first passage with the first stock. Passage notation A and B refers to different trials of the same experiment.

Duplicate analysis of the passage of HLD-2 is shown in Table 3-20. These passages also originated from original passage 1.1 (Table 3-19). Passages 2B and 2A were simultaneously produced from 1.1 with different results. The passage B lineage continued to yield somewhat lower titers than linage A. Both lineages yielded higher titers at passages five and six.

Table 3-20. Duplicate Evaluation of Wild-type HLD-2 Replication for Total Genome RNA Production and RF Production at Weekly Intervals.

Passage and week PI	Total Yield of RT-PCR units per flask	RF/positive strand evaluation End-point	RF/negative strand evaluation End-point	Nested of positive strand reaction	Nested of negative strand reaction
1.1 Wk4	5.7×10^5	10^{-2}	Und[a]	10^{-2}	Und
2B Wk 4	2.9×10^4	Neg	Und	Neg	Und
3B Wk 1	2.8×10^4	Neg	Neg	10^1	Und
Wk 2	1.4×10^5	10^1	10^1	10^1	10^2
Wk 3	2.8×10^5	10^1	10^1	10^1	10^1
Wk 4	1.4×10^6	10^2	10^2	10^2	10^2
4B Wk 1	2.8×10^4	Neg	Neg	Und	Und
Wk 2	2.8×10^5	Neg	Neg	Und	$\geq 10^1$
Wk 3	2.8×10^5	Neg	10^1	$\geq 10^1$	NT[b]
Wk 4	2.8×10^5	10^2	10^2	$\geq 10^2$	10^2
5B Wk4	1.4×10^6	10^3	10^2	10^3	10^2
6B Wk4	2.9×10^6	10^3	10^2	NT	NT

[a] Undiluted [b] Not tested

All of the data indicate that in HLD-2 HAV infection, RF production occurs, is more difficult to detect in early passages, and occurs at a higher level as virus is adapted to cell culture. RNA replication has been suggested as one of the blockages in replication of HAV (Anderson et al., 1998). The irregularity of RF production in HLD-2 infected FRhK-4 cells would fit with this hypothesis. The data we have obtained indicate that culture of HAV is very difficult, which has also been found by many previous investigators (see review by Hollinger and Emerson, 2001). The only clearly successful replication was obtained with a high titer known infectious inoculum and only in one cell line. Continued passage of some of the other cell lines could potentially yield successful replication. This approach of blind passage was used in several of the initial cultures of HAV, since CPE is not apparent (Hollinger and Emerson, 2001). This does not provide a suitable method for detection in wastewater effluent, however, because months of culture are required. A potential approach to the improved culture of wild-type HAV would include use of methods to enhance attachment to the cells, virus entry into the cell, and replication.

Very few investigators have examined HAV infected cells for RF, RI or double-stranded RNA. All studies have used cell culture adapted virus and less sensitive methods available at the time the work was performed. A low level presence of double-stranded RNA in cells infected with a high MOI of HAV in comparison to poliovirus-infected cells has been found (Anderson et al., 1988). Anderson concluded that encapsidation of the positive stranded RNA depleted the pool of RNA available for replication, thus providing the low level of replication in comparison with poliovirus. The more sensitive methods we have used demonstrate RF (double stranded RNA) present in infection with the wild-type virus. In contrast to Anderson's results, we did not find that the pool of RNA available for replication, as measured by the RF present, declined significantly over a four-week period. This may reflect the replication in different cell lines more conducive to replication than the BSC-1 cell line in which his studies were performed. The FRhK-4 cell line was quite permissive for replication of the HLD-2 strain. However we did find a relatively low level of negative strands (RF) in relation to positive strands in the cells infected with a high MOI of cell-culture adapted HAV, approximately a $1:10^5$ ratio (see Figures 3-2 and 3-3).

3.4 Isolation and Replication of Wild-type Enteroviruses, RF Detection and Kinetics

The kinetics of production of RF in cells inoculated with treated wastewater concentrates containing viruses cannot necessarily be predicted from the kinetics of CVB3 production. Therefore, we isolated viruses directly into BGM cells from sewage and used these viruses to understand the kinetics of replication of wild-type viruses that replicate in BGM cells.

Wild-type enteroviruses were isolated from a sewage sample as described in Chapter 2.0. Eleven isolates were obtained by plaque isolation. Each wild-type virus plaque isolate (extracted by chloroform from the agarose sample) was assayed to determine the infectivity titer. Results are shown in Table 3-21. The time required for formation of the plaque assay was longer than required for CVB3 virus, 3-5 days vs. 2 days. The titers ranged from 1×10^4 to 3.8×10^6 PFU/ml. Two of these were identified by sequence analysis of virus from a second passage in BGM cells, #2 (Echovirus 1) and #6 (CVB4).

Table 3-21. Plaque Assay of Wild-type Enterovirus Isolates.

Isolate	Days plaque assay	PFU/ml
WT#1	3days	$3.2x10^5$
WT#2, Echovirus 1	3days	$3.8x10^6$
WT#3	3days	$4.8x10^4$
WT#4	4days	$2.5x10^4$
WT#5	4days	$1.8x10^4$
WT#6, CVB4	5days	$2.4x10^5$
WT#7	4days	$2.3x10^5$
WT#8	4days	$4.4x10^4$
WT#9	5days	$1.1x10^4$
WT#10	4days	$3.0x10^5$
WT#11	4days	$4.4x10^4$

Replication of the equivalent of the second passage of these two identified wild-type viruses, Echovirus 1 and CVB4, was evaluated. BGM cells were infected with 4 PFU/T-25 cm^2 flask by the same method as previously described for CVB3 experiments. Separate flasks were analyzed for total RT-PCR units and for RF production by the column method of RNA extraction. Results of infection with Echovirus 1 are shown in Table 3-22 and results from infection with CVB4 are shown in Table 3-23. Small foci of infection could be seen macroscopically and confirmed microscopically by 36 hr PI in Echo1 infected cells and somewhat later in CVB4 infected cells. By 60 hr PI, 100% of Echo 1 infected cells were rounded and some were even detached and possibly lysed. In contrast the CVB4 infected cells were 100% attached, although rounded with CPE at 72 hr PI. Although the rate of replication was related to the initial titers and time required for isolation by plaque detection, both yielded similar titers upon completion at 48 to 72 hr PI. RF production was clearly evident by 24 hr PI in Echo1 infected cells and 36 hr PI in CVB4 infected cells. This data indicate RF production in wild-type virus infected BGM cells can be detected within two to three days after inoculation with very low levels of virus if replication is efficient in these cells.

Table 3-22. Detection of Wild-type Echovirus 1 and RF in BGM Cells Inoculated with 4 PFU/flask.

Hr Post Infection	Total RT-PCR units	RF/positive strand end-point	RF/negative strand end-point
0	$<1.4x10^1$	Neg	Neg
12	2.8×10^2	10^1	10^1
24	2.8×10^4	10^2	10^3
36	2.8×10^6	10^2	10^3
48	1.4×10^8	10^3	10^3
54	1.4×10^9	10^3	10^3
60	1.4×10^8	10^3	10^3

Table 3-23. Detection of Wild-type CVB4 and RF in BGM Cells Inoculated with 4 PFU/ flask.

Hr Post Infection	Total RT-PCR units	RF/Positive strand end-point	RF/Negative strand end-point
0	1.9×10^1	Und.	Neg.
24	2.8×10^4	10^1	10^1
36	2.8×10^5	10^3	10^3
48	2.9×10^6	10^4	10^4
54	1.4×10^8	10^4	10^4
60	5.8×10^8	10^4	10^5
72	2.9×10^8	10^5	10^5

In the experiments described above, the viral inoculum was diluted in Dulbecco's PBS to achieve the low multiplicity of infection. Experiments have been performed to evaluate the same viruses when the diluent for infection is a wastewater concentrate. Final dilutions of Echovirus 1 and CVB4 were made in a chloroform-extracted wastewater concentrate such that the inoculum volume contained 4 PFU. Cells were harvested after 48 hr PI. Results of this experiment are shown in Table 3-24. When compared with results at 48 hr PI shown in Tables 3-22 and 3-23, production of RF is similar to that obtained when inoculum was diluted in PBS.

Table 3-24. Evaluation of Total Viral RNA and RF Production at 48 hr PI with 4 PFU per T-75 Fflask: Wild-type Echovirus 1 or CVB4 Diluted in Concentrated Treated Wastewater Effluent.

Cells	RT-PCR units	RF, positive sense	RF, negative sense
Echo 1	1.4×10^5	10^3	10^4
CVB4	$\geq 1.4 \times 10^5$	10^5	10^5
Uninfected BGM	$<1 \times 10^1$	Und	Neg

3.5 Evaluation of Wastewater Concentrates for Infectious Virus

All concentrates of wastewater samples were inoculated onto BGM cells, including those prepared by the test methods, in addition to those prepared by the developed method which included the addition of BE for concentration of the treated wastewater. Fourteen different sample dates from Snapfinger were evaluated including a total of 26 different concentration preparations. Fifty-four samples were evaluated from the North Carolina treatment plant with the developed concentration method. Duplicate aliquots of some of the samples were inoculated onto BGM cells for a total of 114 inoculations of BGM T-75 cm^2 flasks of cells. A total of 68 different sample dates from both sites were evaluated. RF RNA was prepared from one-half of each flask by the phenol-chloroform extraction method and assayed for RF with the positive and negative strand assay.

The UNC panenterovirus primers were used to assay all samples initially. Those that were potentially positive based on a light band visualized with the UV light were also evaluated with the CDC non-polio panenterovirus primers. All flasks were examined microscopically daily for presence of toxicity or CPE. While some cells had a slightly altered appearance, most flasks looked like the control cells. All cells were incubated 14 days before harvest. Only one flask had CPE and was harvested four days after inoculation when about 20-30 foci of CPE could be seen in the flask. A second inoculation of this same concentrate did not yield CPE, although it was positive with the UNC primers on the second trial in addition to the first trial. This was the only very strong positive seen in the assay for RF. This isolate has been confirmed by sequence analysis in the CDC enterovirus laboratory of Dr. Oberste. In the 530 bp region sequenced for identity, there was 77.4% identity with CVB3 Nancy strain. This demonstrates that it is a unique isolate. Dr. Oberste confirmed that this percentage variation is similar to that seen in recent clinical isolates. Results of the analysis for RF are shown in Tables 3-27 and 3-28. The samples listed in Table 3-27 have not been confirmed. Most of these collection times are spring or fall dates.

The five confirmed positives represent finding of infectious virus in 7% of the samples from 68 collection dates at the two treatment plants (Table 3-28). If the criteria of clear CPE had been used, only 1.5% of the sampling dates would have been found to have infectious virus. The addition of the RF assay detected more infectious viruses than only CPE determination would detect. Of the five positives, four were concentrated with the developed method and were from the North Carolina site. Only one was from the Snapfinger site and was concentrated without BE addition in the PEG precipitation. In temperate climates, enteroviruses are more commonly observed in the summer and fall seasons (Tani et al., 1995). In warmer climates or coastal climates, viruses may appear year round. The four samples found positive at the North Carolina site were collected in early spring and late fall. This could be explained by the southern near coastal location.

Exact data on the effluent samples from the North Carolina site are not available since they have been archived and it is not feasible to retrieve them. However, necessary criteria were met on all the collection dates. The criteria met included <86 CFU/100 ml of fecal coliforms, residual chlorine <2 mg/L, TSS <30, BOD <20, and ammonia 1-2 mg/L. These criteria meet state requirements. At this facility, the chlorination level is regulated by a flow meter, which varies and is not recorded; therefore, no exact values for chlorination of effluent can be obtained.

Data on the water quality parameters of the Snapfinger samples that were RF positive are shown in Tables 3-25 and 3-26. No unusual data were found on the days of collection of the samples that yielded the RF positive sample (7/3/2000) or on the dates with possible RF positives. There were no permit violations during the four-month period of collections. The geometric mean of fecal coliforms was 400 per month during the time period. There are no data that correlate with the potential presence of infectious enteroviruses.

Table 3-25. Water Quality Parameters on the Selected Dates from the Snapfinger Treatment Plant.

Date	TSS (mg/l)	COD (mg/l)	BOD (mg/l)	Phos (mg/l)	pH		Amm. (mg/l)	Alkalinity (mg/l)	Nitrate-Nitrite (mg/l)	DO Minimum (mg/l)
					min	max				
4/18/2000	3	19	3	0.24	7.0	7.5	<0.1	100	4.87	7.7
5/22/2000	1	23	2	0.36	7.3	7.4	0.6	114	4.80	7.8
6/5/2000	1	20	2	0.15	7.0	7.4	0.9	114	7.63	8.0
7/3/2000	5	14	2	0.58	7/2	7.3	<0.1	106	7.59	6.9

Table 3-26. Coliform Counts and Chlorine Residual at the Snapfinger Treatment Plant.

Date	Total coliforms MPN/100ml	Fecal coliforms MPN/100ml	Chlorine residual before dechlorination (mg/l)	Chlorine residual after dechlorination (mg/l)
4/18/2000	No data	No data	0.54	BDL
5/22/2000	80	<2	0.74	BDL
6/5/2000	8	2	0.73	BDL
7/3/2000	11	<2	0.42	BDL

Table 3-27. Potential Positive Samples, Unconfirmed.

Site and Date	Concentration method	CPE (Trial #)	UNC primers (Trial #)	PE (non-polio) Primers CDC
Snapfinger 4/18/00	8% PEG, 8000 MW, no BE	No	Positive	Negative
Snapfinger 5/22/00	8% PEG, 8000 MW, no BE	No	Positive	Negative
Snapfinger 6/5/00	8% PEG, 8000 MW, no BE	No	Positive (1) Negative (2)	Negative
North Carolina, 9/26/00	8% PEG, 8000 MW, no BE	Possible unusual CPE	Positive	Negative
North Carolina 12/4/00	8% PEG, 8000 MW + BE	No	Positive	Negative
North Carolina 1/16/01	8% PEG, 8000 MW + BE	No	Positive	Negative
North Carolina 2/5/01	8% PEG, 8000 MW + BE	No	Positive	Negative
North Carolina 4/4/01	8% PEG, 8000 MW + BE	No	Positive	Negative
North Carolina 6/6/01	8% PEG, 8000 MW + BE	No	Positive	Negative
North Carolina 9/26/01	8% PEG, 8000 MW + BE	No	Positive (1) Negative (2)	Negative

Table 3-28. Detection of RF and Evidence of Infectious Enteroviruses in Wastewater Concentrates.

Site and Date	Concentration method	CPE (Trial #)	UNC primers (Trial #)	PE (non-polio) Primers CDC
Snapfinger 7/3/00	8% PEG, 6000 MW	No (1) No (2)	Positive (1) Negative (2)	Positive NT
North Carolina 11/27/00	8% PEG, 8000 MW+ BE	No (1) No (2)	Positive (1) Positive (2)	Positive (1) Negative (2)
North Carolina 12/19/00	8% PEG, 8000 MW+ BE	Yes (1) No (2)	Positive (1) Positive (2)	Positive (1) Negative (2)
North Carolina 2/26/01	8% PEG, 8000 MW+ BE	No	Positive	Positive
North Carolina 3/15/01	8% PEG, 8000 MW+ BE	No	Positive	Positive

Confirmation of the one sample that developed CPE was complete in less than a week with the immediate harvest at the first evidence of CPE. With the standard method, more time would have been required to confirm that the sample contained infectious enteroviruses. With the immediate harvest at the first appearance of CPE and confirmation of replication with the molecular method, the infectivity was confirmed rapidly. This developed method offers a definitive assay within a short period of time. If CPE is apparent, the cells can be harvested and confirmation obtained quickly. Confirmation is possible at the end of two weeks if no CPE is apparent. Additional cultures are not necessary. Those viruses that replicate in BGM cells can be detected with the molecular method. Since BGM cells are not the optimum cell line for all viruses present in environmental samples, inoculation in another cell line may provide more robust viral replication and better detection. Molecular confirmation could then be performed on RNA extracts of multiple cell cultures for increased sensitivity of detection of enteroviruses.

CHAPTER 4.0

CONCLUSIONS AND RECOMMENDATIONS

Treated effluent wastewater is frequently discharged into rivers and streams at a point source. These same rivers frequently are the source drinking water for another community; furthermore, rivers near urban areas may be heavily used recreational areas. Viable pathogens remaining in treated effluent can have public health significance. The various treatment strategies remove 90-99.9 % of infectious viruses from sewage. The treated effluent contains inactivated viruses and a varying but much smaller quantity of infectious viruses. Investigators have found infectious viruses present in treated and chlorinated effluent (free residual chlorine 0.2 mg/L) (Bonadonna et al., 2002). Detection of the presence of infectious virus in the effluents is most important. Newer molecular methods are very sensitive and may detect viral nucleic acids from virus particles damaged by the treatment process. These inactivated viruses do not pose a public health threat. This project developed a new method for the detection of infectious virus in treated wastewater effluent.

In this project, we developed a recovery method for infectious virus that yielded a mean recovery of 34% when a model enterovirus (CVB3) was spiked into treated wastewater effluents. Different percentages of PEG with or without the addition of beef extract to the effluent were evaluated to define the optimum method for precipitation of viruses with 8% PEG plus 1.5% BE. Only 10% of the viruses remained in the supernatant, and therefore not all of the inoculated viruses could be recovered in the fractions. The recovery is somewhat lower than recovery achieved from other matrices (Cromeans unpublished). Whether the recovery of the naturally occurring viruses is higher than spiked viruses is not clear; however, our data do suggest this possibility. The initial study on recovery of naturally occurring coliphages with varying percentage of PEG and no BE yielded similar recovery to the optimum recovery of enteroviruses. Recovery of enteroviruses was significantly less without the addition of BE for the precipitation. Therefore the addition of BE was included in recovery of viruses from wastewater collected over one year. Evaluation of the recovery of naturally occurring coliphages with the addition of BE to the PEG precipitation would be useful.

The method for the detection of infectious viruses in treated wastewater was developed by thorough evaluation of laboratory strains of virus. Using the model laboratory virus CVB3, the production of a replicative form (RF) containing a negative strand of RNA complementary to the positive strand of genomic viral RNA could be shown to correlate with production of infectious virus. Extraction of RNA from infected cell cultures, followed by DNase and RNase treatment, yielded an RNA preparation suitable for RT-PCR confirmation of equal quantities of the two strands of RNA.

Kinetics of production of CVB3 in BGM cells demonstrated that with an inoculum of only four infectious units per cell culture flask, replication and confirmation of infectivity was detectable within 24 hr after infection. RF and infectious virus were both detectable until 44 hr

PI, and the quantity of RF remained relative high and constant between 24 and 44 hr PI. In comparison, chlorine disinfected virus did not yield a positive result and no RF production could be detected. These results indicate that only infectious virus can be detected in treated effluent and the excess of chlorine-inactivated viruses potentially present in treated wastewater would not be detected. Chlorine inactivated viruses contained in wastewater concentrates would be detected with only a RT-PCR. The newly developed method was also successful in the detection of replication of a laboratory strain of HAV, HM-175. The HM-175 clone 24A used in the studies can be assayed by plaque assay and therefore cell culture infectivity can be compared with RF production. The quantity of RF and time of appearance was correlated with the quantity of the virus inoculum of a laboratory strain of HAV. A higher infectious inoculum yielded a more rapid production of RF, while a similar level of RF could be obtained with a lower quantity of infectious inoculum and a longer incubation period. The RF remained detectable in the infected cell monolayers for up to one week. The kinetics of production with the different quantity of HAV inoculum is not unexpected. With the low quantity of infectious inoculum, few cells are infected initially and produce RF. As more cells are infected, the quantity of RF increases. This was also observed in the infection with CVB3 at a low MOI.

Culture of wild-type HAV and production of RF was evaluated. Of the four cell lines evaluated, only the FRhK-4 cell line is permissive for culture of wild-type HAV. Only one of three strains of wild-type HAV evaluated in this project could be cultured on this cell line. This isolate has a known primate infectivity titer. About 10^5 animal infectious units were necessary to initiate infection in a flask of 1×10^6 cells. An inoculum of 1000-fold less was unsuccessful. Our data confirm that the current methods for evaluation of HAV in cell culture do not provide accurate information on the infectivity of HAV that may be found in environmental samples such as wastewater effluents. Even in an environmental concentrate, the quantity of virus is unlikely to achieve the concentration of viral particles necessary for initiation of infection under standard conditions. Initiation of infection in cell culture may vary with the particular isolate. Some isolates have not been cultured despite intense efforts. The IC-RT-PCR method can detect intact virus particles from infected cells and was shown to measure the increased production of HAV with successive passages. This method measures only immunologically reactive virus particles, suggesting higher potential for infectivity than an RT-PCR reaction. RF production was irregular in early passages of the HAV wild-type virus, although detectable. Increased quantity of RF was seen and present at earlier times after infection with increasing passage in cell culture, indicating that virus was adapting to cell culture. This phenomenon has been observed in all HAV cell cultures (Hollinger and Emerson, 2001).

Investigators have detected HAV in sewage and final effluent samples from an activated sludge treatment plant by RT-PCR, but were unable to culture virus on BGM cells (Morace et al., 2002). In contrast, other investigators reported culture of HAV from mollusks (DeMedici et al., 2001). Inoculated Frp3 cells (an FRhK-4 derived cell line) were analyzed after 15 days by RT-PCR. However, an increase over quantity of virus initially adsorbed to cells was not shown. Our data demonstrate that wild-type HAV may adsorb to cells and be detected by RT-PCR up to one month after inoculation. This was not followed by a second passage of virus, suggesting that adsorbed virus can be detected but may not initiate infection. Whether HAV adsorption to cells is proof of virus infectivity or capacity to infect humans is not clear. More research is needed to develop methods for detection of infectious HAV. Recently a novel method was described which can distinguish HAV inactivated by high temperature, UV, or chlorine from infectious (noninactivated) HAV (Nuanualsuwan and Cliver, 2002). More research is needed to develop methods for the discrimination of infectious and non-infectious HAV. Addition of components to the culture medium to alter or synchronize replication could improve replication and should be evaluated (Barton and Flanegan, 1997). Capture of the virus with the soluble form of a specific cell receptor followed by RT-PCR could be another approach to the detection of infectious HAV. Recent investigation has demonstrated that a soluble form of HAV cellular receptor can bind and alter HAV particles (Silberstein et al., 2003). Use of one of the soluble forms of the receptor could provide a mechanism to capture HAV that is potentially infectious.

The method developed in this project was best illustrated by the study of two viruses isolated directly from concentrated sewage. Viruses isolated on BGM cells were evaluated at only the second passage for formation of detectable RF in cell culture. RF are detectable at 24 and 36 hr after infection of cultures with only 4 infectious units of Echovirus 1 or CVB4 on T-25 cm^2 monolayers of BGM cells. These time points are about 24 hr before maximum infectious virus production. Cultures inoculated with the same viruses diluted in treated wastewater effluent also initiated production of RF that was detected at 48 hr PI, when CPE was minimal or undetectable. These experiments demonstrate the potential usefulness in detection of wild-type enteroviruses that replicate on BGM cells.

Application of the method to evaluation of treated wastewater concentrates collected during the project confirmed the usefulness of the method. Five of 68 different collection dates contained infectious viruses by the RF detection method, while only one of the five samples produced CPE in BGM cell culture. Only in the culture with CPE were enough viruses produced for sequence analysis by the CDC method. The other four isolates were confirmed by a second set of primers. The detection of potential infectious virus in four samples by the RF method without CPE and at low levels suggests that these were not replicating efficiently in the cells. Classic enterovirus CPE was not observed in these samples. These potentially infectious viruses may replicate more efficiently and possibly with CPE in other cell lines. Although enteroviruses replicate in BGM, some enteroviruses such as coxsackie A viruses replicate better in other cell lines. Expansion of the method to detection of virus replication in multiple cell lines might more readily detect a broader range of infectious virus in treated wastewater effluent and in environmental samples. A study focused on a set of environmental sample concentrates that have been examined by other methods for the presence of virus would be useful. Other methods include the standard method of CPE presence after two passages in culture (a one month assay) or samples that are positive by standard RT-PCR for presence of viral nucleic acid. An additional method includes the integrated cell culture RT-PCR method. Environmental water or wastewater samples that have been determined positive by this method would be good candidates for comparison with the RF method.

The RF method was developed and demonstrated to successfully detect wild-type enteroviruses from sewage. At the present, however, wastewater treatment facilities cannot expect to perform this assay on site. The first part of the assay, however, could be performed with the same equipment and facility as required for the standard method, and then the samples could be forwarded to a laboratory designed to complete the molecular analysis. With a central laboratory for molecular analysis, multiple utilities could participate. Further development of the assay, and new kits that make these assays easier to perform, could make this assay useful in an advanced metropolitan wastewater or water treatment facility. More immediately, the assay should be compared with other assays on collections of environmental samples that have been tested with other methods, such as the standard method and the direct RT-PCR method. This could be done in a research laboratory and would provide confirmation of the usefulness of the method. Although direct RT-PCR of laboratory samples is the most rapid and cost effective method, this does not yield potential infectivity and public health significance. The standard method of two passages in culture is the most costly and has been estimated at $230 for each sample. In contrast the RF method costs about $140 per sample and is complete in about one-half the time required for the standard method.

We assayed inoculated cultures two weeks after inoculation with wastewater concentrates. This requires refeeding the cultures at one week post inoculation, which is the same procedure used in the standard method. This newly developed method should be evaluated with a collection of already characterized samples for comparison of detection at one week after inoculation with the standard two-week time point. If the one-week time point is sufficient, then the total time for the assay is about 9-10 days in comparison with the four weeks required for the standard assay.

In summary, the method we have developed for detection of infectious virus in samples is very effective for wild-type enteroviruses. Very low levels of infectious virus could be detected, as low as four infectious units inoculated onto a T-25 cm^2 flask of susceptible cells. Evidence of infectivity was obtained as soon as six days after infection with a concentrate from secondary treated wastewater. In contrast, the standard method requires two passages in cell culture and therefore two to four weeks to confirm a positive sample

References

Anderson, D.A.; Ross, B.C.; Locarnini, S.A. Restricted replication of hepatitis A virus in cells: encapsidation of viral RNA depletes the pool of RNA available for replication. *J. Virol.* **1988**, 62(11):4201-4206.

Andreoletti, L.; Hober, D.; Becquart, P.; Belaich, S.; Copin, M.; Lambert, V.; Wattre, P. Experimental CVB3-induced chronic myocarditis in two murine strains: evidence of interrelationships between virus replication and myocardial damage in persistent cardiac infection. *J.Med.Virol.* **1997**, 52:206-214.

Barton, D.J.; Flanegan, J.B. Synchronous replication of poliovirus RNA: initiation of negative strand RNA synthesis requires the guanidine-inhibited activity of protein 2C. *J. Virol.* **1997,** 71(11): 8482-8489.

Barton, D.J.; Morasco, B.J.; Flanegan, J.B. Translating ribosomes inhibit poliovirus negative-strand RNA synthesis. *J. Virol.* **1999**, 73:10104-10112.

Bonadonna, L.; Briancesco, R.; Cataldo, C.; Divizia, M.; Donia, D.; Pana, A. Fate of bacterial indicators, viruses and protozoan parasites in a wastewater multi-component treatment system. *Microbiologica* **2002**, 25:413-420.

Brack, K.; Frings, A.; Dotzauer, A.; Vallbracht, A. A cytopathogenic, apoptosis-inducing variant of hepatitis A virus. *J. Virol.* **1998**, 72:3370-3376.

Chapron, C.D.; Ballester, N.A.; Fontaine, J.H.; Frades, C.N.; Margolin, A.B. Detection of astroviruses, enteroviruses, and adenovirus types 40 and 41 in surface waters collected and evaluated by the information collection rule and an integrated cell culture-nested PCR procedure. *Appl. Environ. Microbiol.* **2000,** 66:2520-2525.

Chaves, R.L.; Graff, J.; Normann, A.; Flehmig, B. Specific detection of minus strand hepatitis A virus RNA by tail-PCR following reverse transcription. *Nucleic Acids Res.* **1994,** 22:1919-1920.

Cohen, J.I.; Ticehurst, J.R.; Purcell, R.H.; Buckler-White, A.; Baroudy, B.M. Complete nucleotide sequence of wild-type hepatitis A virus: comparison with different strains of hepatitis A virus and other Picornaviruses. *J. Med. Virol.* **1987,** 61(1):50-59.

Craggs, J.K.; Ball, J.K.; Thomson, B.J.; Irving, W.L.; Grabowska, A.M. Development of a strand-specific RT-PCR based assay to detect the replicative form of hepatitis C virus RNA. *J. Virol. Methods* **2001,** 94:111-120.

Cromeans, T.L.; Sobsey, M.D.; Fields, H.A. Development of a plaque assay for a cytopathic rapidly replicating isolate of hepatitis A virus. *J. Med. Virol.* **1987,** 22:45-56.

Cromeans, T.L.; Fields, H.A.; Sobsey, M.D. Replication kinetics and cytopathic effect of hepatitis A virus. *J. Gen. Virol.* **1989,** 70:2151-2062.

Cromeans, T.L.; Nainan, O.V.; Margolis, H.S. Detection of hepatitis A virus RNA in oyster meat. *Appl. Environ. Microbiol.* **1997,** 63:2460-2463.

Cromeans, T.L.; Favorov, M.O.; Nainan, O.V.; Margolis, H.S. Hepatitis A and E viruses. In *Foodbourne Disease Handbook;* Hui, Y.H.; Sattar, S.A.; Murrell, K.D.; Nip, W.; Stanfield, P.S.; Eds.; 2nd ed; Marcel Dekker, Inc., New York, NY, **2001**, pp 23-76.

Cunningham, L.; Boles, N.E.; Lane, R.J.M.; Dubowitz, V.; Archard, L.C. Persistence of enteroviral RNA in chronic fatigue syndrome is associated with the abnormal production of equal amounts of positive and negative strands of enteroviral RNA. *J. Gen. Virol.* **1990,** 71:1399-1402.

DeMedici, D.; Croci, L.; DiPasquale, S.; Fiore, A.; Toti, L. Detecting the presence of infectious hepatitis A virus in mollusks positive to RT-nested-PCR. *Lett. Appl. Microbiol.* **2001,** 33:362-366.

Deng, M.Y.; Day, S.P.; Cliver, D.O. Detection of hepatitis A virus in environmental samples by antigen-capture PCR. *Appl. Environ. Microbiol.* **1994**, 60(6):1927-1933.

DeSerres, G.; Cromeans, T.L.; Levesque, B.; Brassard, N.; Barthe, C.; Dionne, M.; Prud'homme, H.; Paradis, D.; Shapiro, C.N.; Nainan, O.V.; Margolis, H.S. Molecular confirmation of hepatitis A virus from well water: epidemiology and public health implications. *J. Infect. Diseases* **1999,** 179:37-43.

Divizia, M.; Ruscio, V.; Degener, A.M.; Pana, A. Hepatitis A virus detection in wastewater by PCR and hybridization. *Microbiol. Rev.* **1998,** 21:161-167.

Egger, D.; Pasamontes, L.; Bolten, R.; Boydo, V.; Bienz, K. Reversible dissociation of the poliovirus replication complex: functions and interaction of its components in viral RNA synthesis. *J. Virol.* **1996**, 70:8675-8683.

Egger, D.; Gosert, R.; Bienz, K. Role of cellular structures in viral RNA replication. In *Molecular Biology of Picornaviruses*; Semler, B.L. and Wimmer E. Eds. American Society for Microbiology, Washington D.C., **2002,** pp 247-253.

Frech, B.; Peterhans, E. RT-PCR background priming during reverse transcription. *Nucleic Acids Res.* **1994,** 22:4342-4343.

Gantzer, C.; Maul, A.; Audic, J.M.; Schwartzbrod, L. Detection of infectious enteroviruses, enterovirus genomes, somatic coliphages, and *Bacteroides fragilis* phages in treated wastewater. *Appl. Environ. Microbiol.* **1998,** 64: 4307-4312.

Grabow, W.O.; Botma, K.L.; de Villiers, J.C.; Clay, C.G.; Erasmus, G. Assessment of cell culture and polymerase chain reaction procedures for the detection of polioviruses in wastewater. *Bull World Health Organ*. **1999**, 77:973-980.

Green, D.H.; Lewis, G.D. Enzymatic amplification of enteric viruses from wastewaters. *Water Sci. Tech*. **1995,** 31: 329-336.

Hollinger, J.D.; Emerson, S.V. Hepatitis A Viruses. In *Fields Virology*, 3rd Ed. Vol 1, D.M. Knipe and P.M. Howley.; Eds. Lippincot, Williams, and Wilkins, Philadelphia. **2001**, pp 790-840.

Jansen, R.W.; Newbold, J.E.; Lemon, S.M. Complete nucleotide sequence of a cell culture-adapted variant of hepatitis A virus: comparison with wild-type virus with restricted capacity for in vitro replication. *Virology* **1988**, 163:299-307.

Johnson, K.L.; Sarnow, P. Viral RNA synthesis. In *Human Enterovirus Infections;* Rotbart, H.A.; Ed. American Society for Microbiology, Washington D.C., **1995**, pp 95-112.

Jurgens, C.; Flanegan, J.B. Initiation of poliovirus negative-strand RNA synthesis requires precursor forms of P2 proteins. *J. Virol*. **2003**, 77:1075-1083.

Khanna, B.; Spelbring, J.E.; Innis, B.L.; Robertson, B.H. Characterization of a genetic variant of human hepatitis A virus. *J Med. Virol*. **1992,** 36:118-124.

Krawczynski, N.; Quadri, R.; Rubbia-Brandt, L.; Mondelli, M.; Zarski, J.; Hadengue, A. Detection of genomic and minus-strand of hepatitis C virus RNA in the liver of chronic hepatitis C patients by strand-specific semi-quantitative reverse-transcriptase polymerase chain reaction. *Hepatology* **1999,** 29(2): 536-542.

Lanford, R.E.; Sureau, C.; Jacob, J.R.; White, R.; Fuerst, T. Demonstration of in vitro infection of chimpanzee hepatocytes with hepatitis C virus using strand-specific RT/PCR. *Virology* **1994,** 202: 606-614.

Lanford, R.E.; Chavez, D.; Chisari, F.V.; Sureau, C. Lack of detection of negative-strand hepatitis C virus RNA in peripheral blood mononuclear cells and other extra hepatic tissues by the highly strand-specific rTth reverse transcriptase PCR. *J. Virol*. **1995,** 69: 8079-8083.

Laskus, T.; Radkowski, M.; Wanag, L.; Cianciara, J.; Vargas, H.; Rakeoa, J. Hepatitis C virus negative strand RNA is not detected in peripheral blood mononuclear cells and viral sequences are identical to those in serum: a case against an extra hepatic replication. *J. Gen. Virol*. **1997a,** 78: 2742-2750.

Laskus,T.; Radkowski, M.; Wang, L.; Vargas, H.; Rakela, J. Lack of evidence for hepatitis G virus replication in the livers of patients coinfected with hepatitis C and G viruses. *J. Virol*. **1997b,** 71: 7804-7806.

Laskus,T.; Radkowski, M.; Wang, L.; Vargas, H.; Rakela, J. Detection of hepatitis G virus replication sites by using highly strand-specific Tth-based reverse transcriptase PCR. *J. Virol.* **1998,** 72: 3072-3075.

Lemon, S.M.; Whetter, L.; Chang, K.H.; Brown, E.A. Why do human hepatitis viruses replicate so poorly in cell cultures? *FEMS Microbiol Lett.* **1992**, 79:455-459.

Margolis, H.S.; Nainan, O.V. Identification of virus components in circulating immune complexes isolated during hepatitis A infection. *Hepatology* **1990**, 11:31-37.

McCaustland, K.A.; Bond, W.W.; Bradley, D.W.; Ebert, J.W.; Maynard, J.E. Survival of hepatitis A virus in feces after drying and storage for 1 month. *J. Clin. Microbiol.* **1982,** 16:957-958.

Monpoeho, S.; Maul, A.; Mignotte-Cadiergues, B.; Schwartzbrod, L.; Billaudel, S.; Ferre, V. Best viral elution method available for quantification of enteroviruses in sludge by both cell culture and reverse transcription –PCR. *J. Appl. Environ. Microbiol.* **2001**, 67(6): 2484-2488.

Morace, G., Aulicino, F.A.; Angelozzi, C.; Costanzo, L.; Donadio, F.; Rapicetta, M. Microbial quality of wastewater: detection of hepatitis A virus by reverse transcriptase-polymerase chain reaction. *J. Appl. Microbiol.* **2002,** 92(5):828-836.

Novak, J.E.; Kirkegaard, K. Improved method for detecting poliovirus negative strands used to demonstrate specificity of positive-strand encapsidation and the ratio of positive to negative strands in infected cells. *J. Virol.* **1991,** 65(6): 3384-3387.

Nuanualsuwan, S.; Cliver, D.O. Pretreatment to avoid positive RT-PCR results with inactivated viruses. *J. Virol. Methods* **2002,** 104(2):217-225.

Oberste, M.S.; Maher, K.; Kilpatrick, D.R.; Flemister, M.R.; Brown, B.A.; and Pallansch, M.A. Typing of human enteroviruses by partial sequencing of VP1. *J. Clin. Microbiol.* **1999a,** 37:1288-1293.

Oberste, M.S.; Maher, K.; Kilpatrick, D.R.; and Pallansch, M.A. Molecular evolution of human enteroviruses: Correlation of serotype with VP1 sequence and application to picornavirus classification. **1999b,** *J. Virol.* 73:1941-1948.

Oberste, M.S.; Maher, K.; Flemister, M.R.; Marchetti, G.; Kilpatrick, D.R.; Pallansch, M.A. Comparison of classic and molecular approaches for the identification of untypeable enteroviruses. *J. Clin. Microbiol.* **2000,** 38:1170-1174.

Racaniello, V.R. Picornaviruses: The viruses and their replication. In *Fields Virology;* 4th ed.; Lippincott Williams & Wilkins, Philadelphia, PA. **2001**, Volume 1, pp 685-722.

Reetoo, K.N.; Osman, S.A.; Illavia, S.H.; Cameron-Wilson, C.L.; Banatvala, J.E.; Muir, P. Quantitative analysis of viral RNA kinetics in coxsackievirus B3-induced murine myocarditis: biphasic pattern of clearance following acute infection, with persistence of residual viral RNA throughout and beyond the inflammatory phase of disease. *J. Gen.Virol.* **2000,** 81: 2755-2762.

Reynolds, K.A.; Gerba, C.P.; Pepper, I.L. Detection of infectious enteroviruses by an integrated cell culture-PCR procedure. *Appl. Environ. Microbiol.* **1996**, 62:1424-1427.

Reynolds, K.A.; Roll, K.; Fjuioka, R.S.; Gerba, C.P.; Pepper, I.L. Incidence of enteroviruses in Mamala Bay, Hawaii using cell culture and direct polymerase chain reaction methodologies. *Can. J. Microbiol.* **1998**, 44: 598-604.

Richards, O.C.; Martin, S.C.; Jense, H.G.; Ehrenfeld, E. Structure of poliovirus replicative intermediate RNA. *J. Mol. Biol.***1984**, 173: 325-340.

Schmitz, G.; Dotzauer, A. Proof of hepatitis A virus negative-sense RNA by RNA/DNA-hybrid detection: a method for specific detection of both viral negative-and positive-strand RNA species. *Nucleic Acids Res.* **1998**, 26(22): 5230-5232.

Schwab, K.J.; De Leon, R.; Sobsey, M.D. Concentration and purification of beef extract mock eluates from samples for the detection of enteroviruses, hepatitis A virus, and Norwalk virus by reverse transcription-PCR. *Appl. Environ. Microbiol.* **1995**, 61:531-537.

Silberstein, E.; Xing, L.; van de Beek, W.; Lu, J.; Cheng, H.; and Kaplan, G.G. Alteration of hepatitis A virus (HAV) particles by a soluble form of HAV cellular receptor 1 containing the immunoglobulin- and mucin-like regions. *J. Virol.* **2003**, 77:8765-8774.

Spinner, M.L.; DiGiovanni, G.D. Detection and identification of mammalian reoviruses in surface water by combined cell culture and reverse transcription-PCR. *Appl. Environ. Microbiol.* **2001**, 67(7):3016-3020.

Takeda, Y.; Ashida, M.; Hamada, C. Growth of hepatitis A virus in murine cells. *Acta Virologica* **1996**, 40:201-208.

Tani, N.; Dohi, Y.; Kuramatani, N.; Yonemasu, K. Seasonal distribution of adenovirus, enteroviruses and reoviruses in urban river water. *Microbiol. Immunol.* **1995**, 39(8):577-580.

Teterina, N.L.; Egger, D.; Bienz, K.; Brown, D.M.; Semler, B.L.; Ehrenfeld, E. Requirements for assembly of poliovirus replication complexes and negative–strand RNA synthesis. *J. Virol.* **2001**, 75:3841-3850.

Wheeler, C.M.; Fields, H.A.; Schable, C.A.; Meinke, W.J.; Maynard, J.E. Adsorption, purification, and growth characteristics of hepatitis A virus strain HAS-15 propagated in fetal Rhesus monkey kidney cells. *J. Clin. Microbiol.* **1986**, 434-440.

Yamamoto, K.R.; Alberts, B.M.; Nenzinger, R.; Lawhorne, L.; Treiber, G. Rapid bacteriophage sedimentation in the presence of polyethylene glycol and its application to large-scale virus purification. *Virology* **1970**, 40:734-744.

WASTEWATER UTILITY

Alabama
Montgomery Water Works & Sanitary Sewer Board

Alaska
Anchorage Water & Wastewater Utility

Arizona
Gila Resources

Glendale, City of, Utilities Department

Mesa, City of

Peoria, City of

Pima County Wastewater Management

Arkansas
Little Rock Wastewater Utility

California
Calaveras County Water District

Central Contra Costa Sanitary District

Contra Costa Water District

Crestline Sanitation District

Delta Diablo Sanitation District

Dublin San Ramon Services District

East Bay Dischargers Authority

East Bay Municipal Utility District

El Dorado Irrigation District

Fairfield-Suisun Sewer District

Fresno Department of Public Utilities

Irvine Ranch Water District

Las Virgenes Municipal Water District

Livermore, City of

Lodi, City of

Los Angeles, City of

Los Angeles County, Sanitation Districts of

Napa Sanitation District

Orange County Sanitation District

Palo Alto, City of

Riverside, City of

Sacramento Regional County Sanitation District

San Diego Metropolitan Wastewater Department, City of

San Francisco, City & County of

San Jose, City of

Santa Barbara, City of

Santa Rosa, City of

South Bayside System Authority

South Coast Water District

South Orange County Wastewater Authority

Sunnyvale, City of

Union Sanitary District

West Valley Sanitation District

Colorado
Boulder, City of

Colorado Springs Utilities

Littleton/Englewood Water Pollution Control Plant

Metro Wastewater Reclamation District, Denver

Connecticut
The Mattabasset District

New Haven, City of, WPCA

District of Columbia
District of Columbia Water & Sewer Authority

Florida
Broward, County of

Fort Lauderdale, City of

Gainesville Regional Utilities

JEA

Miami-Dade Water & Sewer Authority

Orange County Utilities Department

Orlando, City of

Reedy Creek Improvement District

Seminole County Environmental Services

St. Petersburg, City of

Stuart Public Utilities

Tallahassee, City of

Tampa, City of

Toho Water Authority

West Palm Beach, City of

Georgia
Atlanta Department of Watershed Management

Augusta, City of

Clayton County Water Authority

Cobb County Water System

Columbus Water Works

Fulton County

Gwinnett County Department of Public Utilities

Savannah, City of

Hawaii
Honolulu, City and County of

Illinois
American Bottoms Wastewater Treatment Plant

Greater Peoria Sanitary District

Kankakee River Metropolitan Agency

Metropolitan Water Reclamation District of Greater Chicago

North Shore Sanitary District

Wheaton Sanitary District

Iowa
Ames, City of

Cedar Rapids Wastewater Facility

Des Moines Metro Wastewater Reclamation Authority

Iowa City, City of

Kansas
Johnson County Unified Wastewater Districts

Unified Government of Wyandotte County/Kansas City, City of

Kentucky
Louisville & Jefferson County Metropolitan Sewer District

Louisiana
Sewerage & Water Board of New Orleans

Maine
Bangor, City of

Portland Water District

Maryland
Anne Arundel County Bureau of Utility Operations

Howard County Department of Public Works

Washington Suburban Sanitary Commission

Massachusetts
Boston Water & Sewer Commission

Upper Blackstone Water Pollution Abatement District

Michigan
Ann Arbor, City of

Detroit, City of

Holland Board of Public Works

Lansing, City of

Owosso Mid-Shiawassee County WWTP

Saginaw, City of

Wayne County Department of Environment

Wyoming, City of

Minnesota
Rochester, City of

Western Lake Superior Sanitary District

Missouri
Independence, City of

Kansas City Missouri Water Services Department

Little Blue Valley Sewer District

Nebraska
Lincoln Wastewater System

Nevada
Henderson, City of

New Jersey
Bergen County Utilities Authority

Passaic Valley Sewerage Commissioners

New York
New York City Department of Environmental Protection

Rockland County Solid Waste Management Authority/ Sewer District

North Carolina
Charlotte/Mecklenburg Utilities

Durham, City of

Metropolitan Sewerage District of Buncombe County

Orange Water & Sewer Authority

Ohio
Akron, City of

Butler County Department of Environmental Services

Columbus, City of

Metropolitan Sewer District of Greater Cincinnati

Northeast Ohio Regional Sewer District

Summit, County of

Oklahoma
Tulsa, City of

Oregon
Clean Water Services

Eugene/Springfield Water Pollution Control

Water Environment Services

Pennsylvania
Philadelphia, City of

University Area Joint Authority, State College

South Carolina
Charleston Commissioners of Public Works

Mount Pleasant Waterworks & Sewer Commission

Spartanburg Sanitary Sewer District

Tennessee
Cleveland, City of

Knoxville Utilities Board

Murfreesboro Water & Sewer Department

Nashville Metro Water Services

Texas
Austin, City of

Dallas Water Utilities

Denton, City of

El Paso Water Utilities

Fort Worth, City of

Gulf Coast Waste Disposal Authority

Houston, City of

San Antonio Water System

Trinity River Authority

WERF SUBSCRIBERS

Utah
Salt Lake City Corporation

Virginia
Alexandria Sanitation Authority
Arlington, County of
Fairfax County Virginia
Hampton Roads Sanitation District
Henrico, County of
Hopewell Regional Wastewater Treatment Facility
Loudoun County Sanitation Authority
Lynchburg Regional WWTP
Prince William County Service Authority
Richmond, City of
Rivanna Water & Sewer Authority

Washington
Edmonds, City of
Everett, City of
King County Department of Natural Resources
Seattle Public Utilities
Sunnyside, Port of
Yakima, City of

Wisconsin
Green Bay Metro Sewerage District
Madison Metropolitan Sewerage District
Milwaukee Metropolitan Sewerage District
Racine, City of
Sheboygan Regional Wastewater Treatment
Wausau Water Works

Australia
South Australian Water Corp.
Sydney Water Corp.
Water Corp. of Western Australia

Canada
Greater Vancouver Regional District
Toronto, City of, Ontario
Winnipeg, City of, Manitoba

Mexico
Servicios de Agua y Drenaje de Monterrey, I.P.D.

New Zealand
Watercare Services Limited

Singapore
Singapore Public Utilities Board

United Kingdom
Yorkshire Water Services Limited

STORMWATER UTILITY

California
Los Angeles, City of, Department of Public Works
Monterey, City of
San Francisco, City & County of
Santa Rosa, City of
Sunnyvale, City of

Colorado
Boulder, City of

Georgia
Griffin, City of

Iowa
Cedar Rapids Wastewater Facility
Des Moines Metro Wastewater Reclamation Authority

Kansas
Overland Park, City of

Kentucky
Louisville & Jefferson County Metropolitan Sewer District

Maine
Portland Water District

Minnesota
Western Lake Superior Sanitary District

North Carolina
Charlotte, City of, Stormwater Services

Pennsylvania
Philadelphia, City of

Tennessee
Chattanooga Storm Water Management

Washington
Seattle Public Utilities

STATE

Arkansas Department of Environmental Quality
Fresno Metropolitan Flood Control District, Calif.
Urban Drainage & Flood Control District, Colo.

CORPORATE

ADS Environmental Services
The ADVENT Group Inc.
Alan Plummer & Associates
Alden Research Laboratory
Aqua-Aerobic Systems, Inc.
Aquateam–Norwegian Water Technology Centre A/S
BaySaver Inc.
BioVir Laboratories, Inc.
Black & Veatch
Boyle Engineering Corporation
Brown & Caldwell
Burns & McDonnell
CABE Associates Inc.
The Cadmus Group
Camp Dresser & McKee Inc.
Carollo Engineers Inc.
Carpenter Environmental Associates Inc.
CDS Technologies Inc.
Chemtrac Systems Inc.
CH2M HILL
Damon S. Williams Associates, LLC
David L. Sheridan, P.C.
Dewling Associates, Inc.
Earth Tech Inc.
Eco-Matrix
Ecovation
Environmental Engineers International
EMA Inc.
The Eshelman Company Inc.
Finkbeiner, Pettis, & Strout Inc.
Freese & Nichols, Inc.
ftn Associates Inc.
Fuss & O'Neill, Inc.
Gannett Fleming Inc.
Golder Associates Inc.
Greeley and Hansen LLC
Hazen & Sawyer, P.C.
HDR Engineering Inc.
HNTB Corporation
HydroQual Inc.
Infilco Degremont Inc.
Ingersoll-Rand Energy Systems
Insituform Technologies Inc.
Institute for Environmental Technology & Industry, Korea
Jacobson Helgoth Consultants Inc.
Jason Consultants Inc.
Jordan, Jones, & Goulding Inc.
KCI Technologies Inc.
Kelly & Weaver, P.C.
Kennedy/Jenks Consultants
Komline Sanderson Engineering Corporation
Lawler, Matusky & Skelly Engineers, LLP
Limno-Tech Inc.
Lombardo Associates Inc.
Malcolm Pirnie Inc.
McKim & Creed
MEC Analytical Systems Inc.
Metcalf & Eddy Inc.
MWH
New England Organics
Odor & Corrosion Technology Consultants Inc. (OCTC)
Oswald Green, LLC
PA Government Services Inc.
Parametrix Inc.
Parsons
Post, Buckley, Schuh & Jernigan
R&D Engineering/Conestoga Rover & Associates
The RETEC Group
RMC, Inc. (Raines, Melton & Carella)
R.M. Towill Corporation
Ross & Associates Ltd.
Royce Technologies
SAIC Maritime Technical Group
Short Elliott Hendrickson, Inc.
Stantec Consulting Group, Inc.
Stormwater Management, Inc.
Synagro Technologies, Inc.
Tetra Tech Inc.
Trojan Technologies Inc.
URS Corporation
Veolia Water NATC
Wade-Trim Inc.
Weston Solutions Inc.
Woodard & Curran
WRc/D&B, LLC
WWETCO, LLC
Zoeller Pump Company

INDUSTRY

American Electric Power
ChevronTexaco Energy Research & Technology Company
The Coca-Cola Company
Dow Chemical Company
DuPont Company
Eastman Kodak Company
Eli Lilly & Company
Merck & Company Inc.
ONDEO Services
Procter & Gamble Company
PSEG Services Corp.
RWE Thames Water Plc
Severn Trent Services Inc.
United Water Services LLC

Note: List as of 7/1/04

WERF Product Order Form

As a benefit of joining the Water Environment Research Foundation, subscribers are entitled to receive one complimentary copy of all final reports and other products.Additional copies are available at cost (usually $10).To order your complimentary copy of a report, please write "free" in the unit price column.WERF keeps track of all orders.If the charge differs from what is shown here, we will call to confirm the total before processing.

Name ____________________ Title ____________________

Organization ____________________

Address ____________________

City __________ State __________ Zip Code __________ Country __________

Phone __________ Fax __________ Email __________

Stock #	Product	Quantity	Unit Price	Total
			Postage & Handling	
			VA Residents Add 4.5% Sales Tax	
			Canadian Residents Add 7% GST	
			TOTAL	

Method of Payment:

(All orders must be prepaid.)

❑ Check or Money Order Enclosed

❑ Visa ❑ Mastercard ❑ American Express

Account No. ____________________ Exp. Date __________

Signature ____________________

Shipping & Handling:

Amount of Order	United States	Canada & Mexico	All Others
Up to but not more than:	Add:	Add:	Add:
$20.00	$5.00*	$8.00	50% of amount
30.00	5.50	8.00	40% of amount
40.00	6.00	8.00	
50.00	6.50	14.00	
60.00	7.00	14.00	
80.00	8.00	14.00	
100.00	10.00	21.00	
150.00	12.50	28.00	
200.00	15.00	35.00	
More than $200.00	Add 20% of order	Add 20% of order	

*minimum amount for all orders

Note: Please make checks payable to the Water Environment Research Foundation.

To Order (Subscribers Only):

Log on to www.werf.org and click on the "Product Catalog."

Phone: (703) 684-2470
Fax: (703) 299-0742.

WERF
Attn: Subscriber Services
635 Slaters Lane
Alexandria, VA 22314-1177

To Order (Non-Subscribers):

Non-subscribers may be able to order WERF publications either through WEF (www.wef.org) or IWAP (www.iwapublishing.com).Visit WERF's website at www.werf.org for details.

IWA Publishing's authorised EU representative for General Product Safety Regulations is Diane D'Arras, 15 rue Duret, 75116 Paris, France, e-mail: safety@iwap.co.uk.

Printed and bound by CPI Group (UK) Ltd, Croydon, CR0 4YY
12/05/2026
02108893-0008